Quality and Total Cost in Buildings and Services Design

Edited by
D J Croome *BSc MSc CEng MInstF MInstP MCIBS FIOA MASHRAE*
and
A F C Sherratt *BSc PhD CEng FIMechE FCIBS MInstR*

The Construction Press
Lancaster, London and New York

This book is based on the conference of the same title, organised in March 1976 by the Construction Industry Conference Centre Ltd. in conjunction with the Royal Institute of British Architects, the Royal Institution of Chartered Surveyors, the Institution of Heating and Ventilating Engineers and the Department of the Environment.

The Construction Press Ltd,
Lancaster, England.

A subsidiary company of Longman Group Ltd,
London. Associated companies, branches and
representatives throughout the world.

Published in the United States of America by
Longman Inc New York.

First published 1977

ISBN 0 904406 40 7

Contents

CONTRIBUTORS

Peter E. Bathurst, FRICS, Regional Works Officer, North West Thames Regional Health Authority. (Organising Committee Member)

F. G. Bennie, AIB, General Manager, Premises Division, National Westminster Bank.

Professor P. Burberry, MSc, ARIBA, CompIHVE, Professor of Building, UMIST, University of Manchester.

Ian Carpenter, M!HVE (representing the Institution of Heating and Ventilating Engineers). Organising Committee Member

M. Carver, BTech(Hons), Engineer, Steenson Varming Mulcahy & Partners, Hemel Hempstead.

J. Michael Cooling, ACGI, BSc(Eng), CEng, FIMech, FIMechE, FIEE, FIHVE, President, the Institution of Heating and Ventilating Engineers, Engineering and Commercial Director, Balfour Kilpatrick Limited.

D. J. Croome, BSc, MSc, CEng, MInstF, MInstP, MCIBS, FIOA, MASHRAE, Senior Lecturer, Dept. of Civil Engineering, University of Technology, Loughborough. Chairman of the Technical Organising Committee.

D. M. Doig, FRICS, President, the Royal Institution of Chartered Surveyors, Senior Partner, Doig & Smith, Glasgow.

Brian Drake, FRICS, Chief Surveyor, Department of Health and Social Security.

John R. J. Ellis, CEng, MIMechE, MIHVE MInstFuel, Partner, Building Design Partnership, Manchester.

Roger Flanagan, MSc, ARICS, AIQS, MIOB, Lecturer, Department of Construction Management, University of Reading.

David Hoar, FRICS, Directing Surveyor, County Architects Department, Nottinghamshire County Council.

Bernard J. Hoskins, CEng, MIMechE, FIPlantE, Engineering Manager, Mullard (Durham).

H. G. Mitchell, MSc(Eng), MIHVE, MASHRAE, Principal, H. G. Mitchell & Partners.

Victor Noble, CEng, MIMechE Dept. of the Environment. (Organising Committee Member)

H. William Pearson, DipArch(B'ham), ARIBA, Partner, Building Design Partnership, Manchester.

J. A. Read, CEng, MICE, Chief Project Manager, W. S. Atkins Group Ltd., Epsom, Surrey.

J. D. M. Robertson, FRICS, MBIM, Principal, The Surveyors Collaborative, Kingston-upon-Thames.

A. F. C. Sherratt, BSc, PhD, CEng, FIMechE, FIHVE, MInstR, Assistant Director and Dean of the Faculty of Architecture and Surveying, Thames Polytechnic, London. (Organising Committee Member)

H. S. Staveley, FRICS, FRSH, FIArb, Senior Partner, Martin Staveley & Partners, Clevedon, Somerset, Chairman of the Building Terotechnology Group.

Henry T. Swain, CBE, FRIBA, AADipl, County Architect, Nottinghamshire County Council.

Nicholas Thompson, DipArch, ARIBA, Partner, Renton Howard Wood Levin Partnership, London.

P. C. Venning FRICS, Partner, Davis, Belfield & Everest.

Professor Douglass Wise, BArch, DipTP, FRIBA, Director, the Institute of Advanced Architectural Studies, University of York.

Preface

In times when the limitations of capital expenditure are the basis for many building decisions, quality is often forgotten. In a world where countries equate success with monetary acquisition, and where a million pounds becomes insignificant in a land deal, real values are too often forgotten by Government, Local Authorities, professionals and individuals alike.

The false deity of money is not the only problem; keeping abreast of change is another. Man's materials, energy, space and financial resources alter; styles and fashions vary too. Our social, cultural and living needs in one decade change their emphasis and colour in another. But buildings are long-term investments and, in consequence, designers need to show foresight and flexibility in their approach if the buildings they design are to be useful in successive decades, thus making them an enviable heritage. So much of life is spent inside buildings and so much of our urban and city landscapes are buildings, that they must in their concept be related to people. Buildings should enrich our society.

This book (and the conference on which it was based) arose from the belief that it is wiser to consider the long-term costs of designing, constructing and operating buildings than just the initial construction costs. But in practice, not only the level of the capital budget set before a building is conceived, but the allocation of this budget, based on precedent is often defined too—a system which places constraints on the designer which in practice can have the effect of making quality and cost control incompatible objectives.

The fact that these objectives are not necessarily incompatible is adequately demonstrated by the contents of this book. The team of specialist contributors have set out in concise terms just what is meant by quality and cost and they explain how constructing buildings of a quality that will satisfy the client for an extended period of time is not impossible when working within the bounds of a controlled budget. This is demonstrated by case studies of three successful long-term buildings which were nevertheless built within the limitations of tight cost controls.

There is much to learn from the considerable American experience in relating building performance and cost control. This is discussed in a special chapter covering current developments in the United States.

The contents of this book provide a framework for developing design criteria which will ensure that buildings can still be constructed within the limitations of tight cost control, yet of a quality that will provide for long-term client satisfaction. In a world where a low capital cost philosophy generally prevails, achieving this objective must now be one of the main considerations in the design of new buildings.

D. J. Croome, Loughborough.
A. F. C. Sherratt, London.

ACKNOWLEDGEMENTS

The editors and technical organising committee wish to thank all the people who have participated in the arrangement and operation of the conference and in the production of this book.

A particular word of thanks is given to Mrs. Diana Bell and her colleagues at The Construction Industry Conference Centre for their efficiency in the organisation and administration.

Chapter 1

The Philosophy of Value in Building Design & Use

F. G. Bennie

Cato said, over 2000 years ago,"*Do not buy what you want, but what you need; what you do not need is dear at a farthing*". This statement is as true today as it was so many years ago. Essentially it is a simple statement and one that should be fully comprehended by all clients and the professionals who advise them. So often, however, it is overlooked by the smaller and less sophisticated client and I suspect also that this principle is at times overlooked by some of the larger clients.

It is perhaps felt that bankers have a special eye for value and making the most of resources when balancing quality with cost. Looking back over many contracts and at many buildings, I know that historically this has not always been so, although there are many buildings which were designed as banks some 150 years or more ago and where the same basic structure, albeit modernised internally, supports the same function for which the building was originally designed. Surprisingly the cost of maintenance over the years of some of these buildings, mainly stone structures, is not as great as one might suppose, and if it were not for changing ideas of shop presentation, would still be suitable; in fact, in many areas these stone-built properties add some character to an otherwise commercial environment.

Basically the need for a building is related to the activities required within the building. Thus the function of the individual, the body of people or the machine to be housed within the building should determine the form of construction. Corbusier acknowledged this with his dictum that "*a house is a machine for living in* "; he was asserting that design should suit purpose. The cost of fulfilling the need to be satisfied will require a careful examination of the financial implications, for all capital expenditure should generate a return and that return should be either that percentage on capital employed which is acceptable within the established criteria of the organisation or business, or in the case of public buildings—civic gain; but in the latter also I believe that there are cost criteria capable of evaluation. It is not, I think, sensible to approach the problem from any other direction. I do not believe that one should first decide what money is available and then see how best it can be spent, but, do not overlook the fact that cost is a major consideration.

SPECIFIC CONSIDERATIONS FOR A NEW OFFICE BUILDING

When we embark upon a new project or development we carefully consider with our own professionals our requirements before giving a brief to the architect. The following points are some of the main considerations. (They are not listed necessarily in order of importance or priority.)

1. It must be suitable for its purpose.

2. Bearing function in mind the design must be economical of construction.

3. It must be capable of providing a reasonable working environment.

4. Externally the building should be attractive to the occupants and to the community at large. Internally it must be attractive to the occupants.

5. It must be easily maintained and running costs should not be excessive.

6. The initial building cost should be reasonable.

7. The building should be occupiable in a reasonable minimum construction time.

8. In large complexes personnel must be able to get to their work station quickly.

It follows that due regard has to be taken of the economic climate, the cost of money, the availability of labour, the state of the building industry and the economic use of scarce resources. It goes without saying, of course, that if it is a building not required for owner occupation or for part owner occupation, then the letting potential is a consideration. I will now enlarge on these points in greater detail.

1. Suitability for its purpose

This may appear to be stating the obvious, but it is surprising how external appearances, planning, or site dictates inhibit the shape, size and convenience of the interior function and thereby the efficiency of the work-force. What I aim to achieve is a building envelope for the function, which will produce working and service areas of usable shape, and that by and large expenditure is not lavished on non-functional architectural excesses (i.e. departures from standard components: windows, metal sections, etc.).

2. Economical Design

To me this means a building of such depth and floor to ceiling height that will permit the maximum daylight penetration and full utilisation of space. This depth should also be such as will provide an area capable, if need be, of being broken down into individual offices without waste.

When Sir George Gilbert Scott showed Lord Palmerston his plans for the new Foreign Office, the Prime Minister objected that "It would look too much like a Continental Cathedral and would be too dark for my clerks to work in." The plans were later adopted for St. Pancras Station where, presumably, the clerks had better eyesight. It is now a protected building!

In other words, working within the building regulations we have to provide the most efficient *gross/net* ratios and position staircases and lift cars, ducts, etc., so as to give the maximum use of space and flexibility of working areas.

3. Working Environment

(a) *Lighting*. At times I think some of us have been encouraged by experts to over-illuminate. Insufficient regard is taken of the height and nature of the work surface, and the texture and colour of the surrounding wall and ceiling surfaces. This in turn leads on to excess running and maintenance costs, which may only be justifiable where a heat reclaim system is in operation, and to higher replacement costs in addition to the higher initial cost.

(b) *Heating and Ventilating*. Too much heat is as much an evil as too little, and both lead to inefficiency, bad humour and discomfort. An adequacy of fresh air is also important. In major city centres it has become the accepted fashion to provide full air conditioning because of noise, dirt and excessive exhaust fumes. In many locations, such sophistication is essential. Similarly, it is necessary for some other reasons as well to apply this dictum to high rise buildings. There are situations, however, even in city centres where I question the necessity for full air conditioning. Certainly there are many locations away from city centres where I would not provide full air conditioning and we have found our staff appreciate being able to open windows.

4. Attractive Appearances

I believe that the general public are very conscious of the buildings around them in general terms, and are quick to dislike anything out of scale or out of tune with the street or neighbourhood, and equally quick to dislike a building that is allowed to fall into disrepair. A responsible developer should be aware of this and should, through his architect, seek to provide buildings which are of good appearance initially and because of intelligent use of materials and good detailing will take their good looks into old age if properly maintained.

2

The use of good materials is so important. Architects should remember that many of their clients are people who build only once or twice in their lifetime and, unlike many of the large commercial or public clients, they know little of the nature of materials and are mostly impressed by appearance. Architects have, in my view, a responsibility to advise a client on the best alternative materials available and should see that adequate advice is tendered on care and maintenance. It is not good enough for a material to be used which in one environment is perfect but in another is disastrous. I have in mind some of the marbles which are excellent internal materials but bad when used externally, e.g. Swedish Green, which when exposed to full sun bleaches badly and rapidly loses colour.

I find also a great awareness amongst staff occupying buildings of the external appearance, and certainly a very keen awareness of the interiors. Although very difficult to quantify, there is some evidence that people work better and keep better hours given a pleasant and rewarding environment.

5. Easy Maintenance and Reasonable Running Costs

Once a building comes into occupation the initial capital outlay is completed. The cost of the building does not, however, end here, for cleaning, heating, cooling, lighting, decorating, repairing and maintaining all continue during the life of the building. They can be items of considerable significance, and the design team must give due consideration to their implications, and it is here that true value should be judged and balanced against revenue, the tax situation, availability of labour, and rates of inflation.

Is it better to spend more now at the gain of a lower maintenance charge, and does a reduction in quality and a saving in outlay really mean economy over a reasonable term? Are the concrete and largely uncleanable structures of today really an improvement financially on the gently mellowing brick and stone of former years?

A typical example within banks is the way in which we have moved from marble and woodblock flooring through rubber and linoleum to carpet. The latter is now no more expensive to install, is much easier to replace when worn, does not so readily show up minor defects in the sub-floor, and because every charwoman knows how to keep it clean, keeps a better appearance during its life. We are, however, experiencing great problems in keeping the other types of floor in good and clean condition without excessive recourse to outside contractors.

Carpet research carried out some years ago proved that carpet, when viewed against its initial cost and its ultimate up-keep costs, was a better buy than most other flooring materials. Perhaps the Building Research Establishment should evaluate initial cost against costs-in-use, and publish "best buy" recommendations?

Much can be done to save fuel costs by a careful study of maximum demand and the selection of the proper tariff. In this regard we have used the services of outside specialists who have saved us many thousands of pounds.

6. Initial Building Cost

In National Westminster Bank we work within a budgeting system whereby we are projecting forward anticipated expenditure which has, in total, a relationship to the availability of total Group resources. Most Companies and Departments work within such disciplines and obviously capital cost of a project is a major consideration. It must not, however, be seen in isolation and must relate, in my view, to maintenance and running costs so that the totality of a building's cost is seen.

7. Occupation Time

A new project can tie up a lot of capital for a long time during which no interest is earned. It is, therefore, essential that a new project generates an income, notional or real, in the shortest reasonable time scale. The design can have a considerable influence upon this aspect, and here, I am afraid, the planners and conservationists can have a serious delaying effect upon the early commencement of many projects. It is necessary that there shall be adequate pre-planning and adequate and early design details available for the contractors. Materials and components selected should be readily available, sub-contracts placed in good time, and on occasions, because of the complexities or difficulties of a project, the early nomination of a main contractor whose expertise is available at the early stages of planning could be of great benefit in optimising the construction techniques and organisation of the project.

8. Movement of Personnel

The siting of entrances, staircase, lifts, escalators, toilets and cloakrooms to the main work stations is very important. Bottle-necks and tortuous routes are to be avoided, and fast, efficient and capacious lifts are an important consideration, particularly in high rise buildings. I find that from my experience we tend to under-provide lift capacity; insufficient margin seems also to be built in for down time, which seems all too frequent these days. In buildings of high personnel occupancy by one occupier I believe more use could be made of the Paternoster type of lift. We have installed one of these machines in one of our eight storey buildings in Manchester with much success. The staff use them without problems and they have given us five years trouble-free use to date. The down time for repairs is negligible and they have cut down considerably time wasted walking up stairs or waiting for lifts, and represent a prime example of initial cost being repaid in a reduction in staff time wasted.

Perfect design with all amenities, easy to run, easily maintained, with a perfect environment, quickly erected and forming a very efficient and flexible office or work place.

COST

Our objective therefore is a perfect design, providing all amenities, easy to run and maintain and having a perfect environment, quickly erected and forming a very efficient and flexible office or work place. But at what cost?

What do we mean by cost? Is it the initial cost?—the running cost?-the maintenance cost?—or, is it a combination of some or all of them? All these items should be considered before final commitment and the answer will not be the same for every developer. It will in most cases be a compromise. In order to arrive at the point of decision I believe we should seek an elemental cost plan which put forward the various alternatives whether materials, components and equipment, and including, where appropriate, the estimated annual maintenance cost, the running costs, the life of the material or equipment and the incidental replacement costs. Initial capital costs for conserving energy are now very relevant and show a very rewarding influence upon cost in use.

How many of us equate the total package? What mileage do we get out of our initial expenditure? After adding the cost of site acquisition to the capital costs, the fees and the cost of capital, do we then evaluate the running costs per annum over the next 20 years or so? This surely is as important as the cost of capital? Of course, as an owner occupier, it is easy to talk in this way, for I have a vested interest in the long term life of most buildings I construct. In other words, the totality of cost falls upon me in one way or another and I am in the final analysis considering the availability of capital against the overall earnings or revenue position. In this context it should not be forgotten that in a commercial organisation a charge against revenue received is effectively a charge against corporation tax at 52%. Circumstances, therefore, might decide that higher revenue charges were preferable to higher capital costs.

I often wonder how much of the philosophy of total cost in building is considered by the developer who is building only for market occupation. Bearing in mind that so many of these developers are in normal times seeking to find one tenant for these buildings on long lease subject to frequent rent review, then the full repairing responsibility is transferred to the tenant. Do tenants consider sufficiently the building specification or its impact on them when agreeing the rentals? When one considers the degree of sophistication in modern buildings and the 15/20 year life-span of most mechanical equipment, this aspect is one of considerable moment.

There are too many buildings where massive items of plant are placed in positions in the basement or sub-basement without any thought having been given to the position which will apply when replacement is necessary.

We all realise only too well the high cost of building today and attention must be drawn to the impact which the Planning Authorities are having on projects. Much is heard these days of the social demands of architecture, but I do not believe that planners hold the peoples' conscience. They are implementers of government policy, albeit with personal and local prejudices. Government policy may reflect social demands or it may ignore majority views in favour of ideological directives. It is certainly affected by vociferous minority groups, and all too often solutions are imposed in the name of the people but with little reference to the ordinary person. Some of the planners' solutions dictate uneconomic terms which invalidate the majority of the arguments of economic balance, in particular high rise office blocks which are difficult and costly to construct and often wasteful in terms of occupation costs. Delay in quite simple issues can be of such proportion that at the end of the day and after, in some cases, lengthy appeal, the viability of a scheme has disappeared and very often the community at large also will have lost.

Thus, the client wanting to erect a building for a particular purpose is immediately constrained by plot ratio, height guide lines, building lines, angles of light, etc., even before the professional team gets to grips with the project.

Thus, can it be seen that compromise is with us from the beginning and the near perfect building that seemed to flow from my philosophies is thereby affected. This compromise must, of course, range over all the disciplines involved: *financial; design; lettable or usable area to core area ratios; servicing of the building and services to the building, including primary fuel.*

These all have an effect on the maintenance and operating costs of the building, and this seems to me to be the main issue, namely, the discussion of *effective* building costs.

THE PROFESSIONAL TEAM—ITS VALUE TO THE CLIENT

Historically many of the world's great buildings were the work of one man, the architect, who conceived, designed and built. He understood the basic principles of structures and when he moved into new *ideas* he very often learnt the hard way and very often to the cost of his patron. This has not changed as is well illustrated by Ronan Point; box girder bridges; high alumina cement?

We now have such sophistication and complicated structural problems emanating from design that there are three professions involved in the design alone. This of itself would not be so bad if they always worked together to the common benefit of their client; it seems that too often they are so jealous of their own professional persuasion that their efforts are almost counter productive.

Not long ago simple commercial buildings were designed by an architect without the help of engineers or mechanical services engineers; the engineer only being required on occasions to advise on foundations. The architect would have been quite happy to design the structure and obtain the necessary consents. Now he seems incapable or reluctant to do so; when I question this I am told it is because of the requirements of the Local Authorities that they see the structural calculations. Is not structure still part of an architect's training?

Perhaps one should ask the question: do we make things too complicated? Certainly today we have to build on sites and soils hitherto considered totally unsuitable and, no doubt, perhaps unthinkingly, we ask for environmental conditions and sophisticated control of them which is not truly justified, but are we sure the extra expertise we pay for in fees really compensates every time in building costs and more efficient services compared with the old rule of thumb methods, or are we locked into an ever increasingly complicated system involving more and more specialists because Regulations and *keeping up with the Joneses* demands it?

Cost can be badly affected by poor co-ordination of the professional skills and by poor project control. Historically again this has largely been the function of the architect, and I would not for one moment suggest that there are not some architects today who are well aware of this function and carry it out with great efficiency. I am bound to observe, however, that from my country-wide experience there are still too many who are in this respect quite inept. It seems to be an aspect requiring greater attention when young people are being trained.

In these several ways costs are increased and value affected. I find myself much attracted to the view that the professions should come together in multi-professional practices where the necessary skills are closely co-ordinated and management is part of that skill. I have recently used such a practice on two of my projects and I can only say that I have been very impressed, particularly with their planning, cost control, and project management.

VALUE IN MATERIALS AND DESIGN

I have already commented on over fussy design and the use of non-standard sections, etc. One of the virtues which I have seen in America is the architect's discipline so far as the use of standard materials is concerned. I refer particularly to doors, architraves, metal work, stone and marble. He will work with these constraints and with sufficient attention to detail to produce a building of quality.

Over the last two decades we have seen many examples of poor detailing, particularly externally where good materials have been used, but because of lack of thought or understanding of the materials the detail has resulted in bad weathering or discoloration? There are, of course, some inescapable difficulties in the use of standard parts which are likely to need replacing in the course of the life of the building. For example many firms that now make vital parts are unlikely to be in existence in 20 or 30 years' time—but even if they are, it is unlikely that they will still be producing the same components or ones which are compatible with those fitted today.

Time is money and it is therefore essential that the design team keep this closely in mind. The speed of erection is a most important consideration and a simple structural frame plays a vital part. Once again we come to the element of choice. As examples:

1.   Reinforced concrete, structural steel or combination of both.

2.   Elemental cladding, pre-cast cladding.

3.   A building system.

4.   Dry internal finishes, wet finishes.

Whatever the choice, the cost is different and the time scale different. As an example of this consideration, in a 550,000 sq ft development at Leman Street on the fringe of the City, which will house some 3000 staff and include a major computer complex, we required that there should be a high degree of flexibility. We decided, therefore, to build on the basis of a structural grid of 30 ′ × 60 ′; and to eliminate floor screeds, floor ducts and floor finishes by the installation of a computer type suspended floor thus eliminating trades and giving maximum flexibility for floor layouts. This is, I feel, a valuable example of the possibility of higher initial cost giving considerable long term savings.

When I see some of the fine buildings erected in the late 20's and early 30's that have been demolished to make way for new buildings, some of which might well produce better working environments but which do not in any way measure up to the building quality of their predecessors, I shudder at the enormity of the wastage this represents. I am certain that when they were built the architect and client did not envisage a life span of less than 80/100 years. Nevertheless, it has been seen to be more economic to demolish and rebuild than to attempt to alter, refurbish and provide air conditioning.

Most modern buildings, not only by tradition but principally by regulation, have to be built with a basic structure so strong that they can last at least 100 years, and because it is possible to clad the structure inside and out with materials of various leases of life, do we think of cladding or covering these structures with a view to 'Heritage' architecture or our pockets? In past times up to, say, 40 years ago, the structure was, in most cases, integral with the cladding and the problem did not arise, but today we have a choice. Do the social demands of architecture require an ability to reface structures every 30 or 40 years, or do they require us to build in our own generation style for 'Heritage'?

What problems then do we store up for the future if the present trend towards greater conservation continues. It could be said hardly to be in the best interests of property owners to encourage architects to provide buildings of singular or impressive design, if by so doing they are not in the future, in control of their destiny.

Alex Gordon's long-life, loose-fit, low-energy building principle where extra money may be spent initially on making a building adaptable as the needs of its occupier change has much to commend it. We have, of course, a long way to go in this regard but this philosophy has motivated us in the design of the Leman Street project.

QUALITY

The most widely accepted definition of quality is *degree of excellence* but many other definitions exist.

Difficulties in defining quality in building stem perhaps from unconscious confusion in people's minds between architecture (the art of the built environment) and building. We all have preconceived ideas about how certain buildings should look. For example, Town Halls: impressive, Banks: dignified, Churches: vertical in concept and echoing, Schools: open and spacious, Houses: comfortable. We also carry preconceived ideas about quality in terms of material. Marble, bronze and stone all conjure up grandeur and subconsciously quality. Size and striking shape are always impressive and may be of quality, but not always.

Quality in a building is, I consider, a combination of many things and quality can be external or internal or both and, while I could not disagree that money helps, a building of quality does not have to be lavishly expensive.

Much has been said in recent years of the quality of life; it is a term which was much used by government and by commercial organisations to illustrate the benefits of decentralisation away from the Greater London area. It embraces commuting conditions, improved housing at lower cost, less crowded schools, the easy access to countryside, seaside and leisure activities, but above all it referred to the much improved new factories or offices that staff were moved into and to the improved staff facilities within factory or office that employers could provide in areas of much lower cost. The quality of life is an all embracing term—it refers to all those areas of concern to us in our living and working environment.

During their working lives, workers spend just over half their waking hours at their place of work. Clearly, therefore, environment is an important consideration. As important is the provision of welfare and leisure facilities; too often we see a rigid adherence to the minimum requirements of legislation whereas I believe that it is the duty of every good employer to provide the best working conditions, subject only to reasonable cost constraints.

Quality can in all things be a very subjective judgement. I see quality as referring not only to the professional's view of the building's design and technical aspects, but the way in which the interior is seen by those who work within it, and the exterior by the world at large. The exterior of a building cannot, of course, be entirely divorced from its surroundings, and landscaping can have a not inconsiderable bearing on exterior quality. There can be no one arbiter of quality, for it is the summation of the input of several people and the judgement of many more. We should all strive to achieve it. It does not always emerge only from the use of high cost materials: it is a combination of careful thought, evaluation and the use of good detailing, the right material, colour and texture.

In an attempt to summarize all these thoughts, a building of quality is one that is effective; one that is right for its purpose; right in scale and proportion; right in solid to void; right in colour and texture; and right in the environmental conditions it provides. All in all it is a place in which to work that more nearly equates with a place in which to live.

When finished we should all ask ourselves whether we could be proud to own it and pleased to work in it. If, truthfully, we can answer 'yes' we have probably achieved success. If the answer is 'no', we must find out why and learn from our failure.

CONCLUSION

If I may have created the impression that the client is perfect and the professionals far less so, then I hasten to correct this, there are as many poor clients as poor professionals. Viewed as a whole the construction industry and all who are employed in it has still a long way to go before it can be said to be effective, efficient, and economic. In my view the client has a very real part to play, with the architects, engineers, cost consultants, surveyors and builders to ensure that at the end of the day his money has been well spent. He must consider the total cost, not the initial cost, and aim for reasonable quality internally and externally.

Thus my philosophy of building is based upon a gathering together of judgements from the many facets of the professional team. Then as an involved client, who is ultimately footing the bill, to listen to and decide from the arguments presented which path to follow so that an enlightened and effective project results and a building created which we are proud to own and pleased to work in.

DISCUSSION

H. S. Robinson (National Westminster Bank Ltd., Premises Division)

From Mr Bennie's paper it would appear that as far as expenditure on bank buildings is concerned the sky is the limit. Would Mr Bennie like to comment on this?

F. G. Bennie

I am afraid that would be a totally erroneous supposition, banks have to be as cost conscious as other commercial organisations. We carry out capital appraisal exercises on every investment we make. Whether it is a bank building or buying another company, or whether it is buying into some foreign country, by making some basic assumptions we carry out a complete capital appraisal carefully quantifying our cost of capital where at the moment we use a cost of $14\frac{1}{2}\%$. Any proposition put forward to our board must be on the basis that it satisfies that cost of capital or that there are very very good management judgements and management reasons why exceptionally we should depart from cost of capital criteria. An example of the latter might be our new head office building in Bishopsgate, 183 m (600 ft) high building built on London clay—higher than has ever been attempted before and in my judgement far too high. Nevertheless, it will be a head office building of an international organisation. The cost is inevitably higher than would be applied to normal projects. By adhering to the same disciplines as others our buildings compare more than favourably with any other developers standards and costs.

A. J. Colledge (Department of the Environment)

Mr Bennie referred to the use of a computer type floor, because of the flexibility it gave in floor layout, he instanced this idea as a valuable example of the possibility of higher initial cost giving a considerable long term saving. This is contrary to our experience at the Department of the Environment. We found the computer type floor to be a very very expensive initial cost which we could not justify in a number of cases. We also experienced difficulty with the Greater London Council on the question of fire risk. Whilst acceptable in the comparatively limited area of a computer room, application to the whole floor area of a very large office block is quite another matter. The objections on the grounds of fire risk are very serious ones.

F. G. Bennie

The comment was made in the particular context of a building we are putting up for our own purposes. It is a large configuration—about 56 000m²—including a major computer building, paper processing plant with very heavy reader sorters, a great deal of automated machinery within the building generally and a large office block which will also contain machinery of various types including electric typewriters, printers and coders. In this kind of building, if you have large open floor areas, and a grid of 18.3 m × 9.1 m you never can have enough electrical points, you never can have your outlets where you want them because if you put them in what you think is the right place to start with, then the first move you make in the building you find that they are all in the wrong place, you have no flexibility. There is an enormous amount of wiring in this complex which we costed very carefully. With the flexibility which we required we found that to have worked to a conventional pattern of screeded floors with trunking, we would have had to compromise our grid by having intermediate cores running up the building. The costings showed that in the particular circumstances suspended floors gave us greater economy.

We did look at this type of solution in the context of our Bishopsgate Tower. Would it not have been simpler to have put in a suspended floor everywhere? We could have satisfied the fire regulations by installing a suitable fire protection system eg sprinklers but it was not found to be viable financially.

D. M. Andrew (National Westminster Bank)

On the cost effectiveness of suspended floors in intensively used office areas—I have an experience where some 1100m² of false flooring is actively in use in a large data preparation and clerical control area in the Southend area and this has been invaluable—it is a modified suspended floor working on a depth of only 100 mm and with a 3 metre liftable tray module—the reason why it is so useful is because of flexibility; speed of change of electrical and telephone circuits matter to us a great deal.

R. Cullen (Architects Design Group)

I am sorry that my dream has been shattered and that I can no longer persuade my clients to save money by using expensive finishes, but I have every intention of going on attempting to persuade my clients to use expensive finishes because instinctively I feel that it is the right thing to do. Surely we can use good materials if they are available and I cannot see really that they need necessarily place a burden on our balance of payments. A wide range of materials including natural materials are available and they do not necessarily have to be imported. So I do not see that the financial exercise in one sense will give us the wrong financial result in another sense ie the national economy.

My blood pressure rose immediately when I was told about planners and the delays that they cause. I am a conservationist. I like delays and I do not like 180 m high buildings, I never have done except in very special situations. There is a social cost to high rise and when we are looking at total cost of buildings should we not include an assessment of the social cost. Also we have talked about cost benefits of 'good' building, should we not evaluate the 'cost loss' of bad building.

F. G. Bennie

The scarcity of resources in this small island has been mentioned and surely one of the scarcest resources, if we are going to carefully plan and balance the use of our land, is land itself. If we equate that with all the other constraints which we have imposed upon ourselves and they are all self imposed rules, rights of light, building regulations, building lines etc. I do not believe it is possible any longer, in urban situations, to think only in terms of low rise buildings. I would agree with Mr Cullen in his abhorence of high rise buildings as an environment for people to live in, but when it comes to offices, provided they are reasonably well spaced and not crowded together and provided they are not a solution imposed by planning constraint, I believe it is possible to have high rise building that is not necessarily detrimental to the environment as a whole.

H. T. Swain (Nottinghamshire County Council)

I am very glad Mr Cullen made that contribution speaking as an architect. I think you should have figures, but you do not necessarily have to follow the figures and I share Mr Cullen's theory that we should have the best possible materials in our buildings. Perhaps I only differ with him in saying that I should try to persuade my clients to use the best materials because they are the best materials, I will not try to substantiate them in terms of running cost. I think Mr. Cullen is absolutely right that we have to raise quality, from the point of view of the quality of the building itself, and the users of the building.

One of the things that has not been referred to here is the awful business of grinding money out of Local Authority committees for maintenance. Somehow, we have no problems in getting capital costs of several million pounds which go through the county committees at the drop of a hat for comprehensive schools etc. but then each year the battle starts for getting enough money to repaint buildings. Quite apart from total cost in use, speaking as a County Architect, I like buildings where I do not have to spend too much money on building maintenance because it is an endless battle to get people to write a cheque each year for maintaining buildings, even if you convince them the sum is a small one.

F. G. Bennie

I would just like to add and I hope that Mr Cullen as an architect will accept this, that I do get just a little worried at times when the conservation lobby gets at us to preserve everything, when I do sincerely believe that every generation has a right to give something to posterity. When we look back over the history of this country we have some marvellous buildings which identify themselves with various ages of architecture. If we are not very careful, with planning and conservation going as it is, will we have anything to leave to posterity?

M. Curtis (University of Reading)

Mr Bennie made a very fashionable swipe at air conditioning and asked do we need air conditioning? We want air conditioning because it improves the internal environment; if we do it properly. He said we had heating and ventilation right some years ago, we had vertical transport right some years ago, we used stairs. Mr Bennie is now prepared to use a lift. We had domestic heating right some years ago, we used open fires. We have now gone to domestic central heating. Our expectations of the internal environment have changed and we are expecting a better environment inside offices.

F. G. Bennie

I do not think I said that the old fashioned methods of heating were always right, what I was trying to say is that we seem to have a fetish, today, that because in certain situations something is right, then in all situations it is right. I built what was probably one of the first air conditioned office buildings in this country in the late 1950's. It was built on the edge of a new town in a very lovely situation. It was a building which, in terms of

9

general environment, offered far far more than anything we had in London. It was fashionable at that time for the informed developer to think in terms of air conditioning. The problems we inherited were enormous.

Since then, we have built many more air conditioned buildings and it is a peculiar thing that very very rarely have I built and occupied an air conditioned building where I have satisfied more than 40 to 50% of the total inhabitants. I do not know the experiences of other owner/occupiers or other professionals that have to live with the buildings they put up because they are in a continuing relationship, it is very very difficult even with the finest air conditioning throughout the building, to get the humidity right, the temperature right, to put in enough controls to satisfy everyone. The one thing we are finding more and more is that people working within our buildings, want a say in the environment in which they live; they want to be able to control that environment.

Whatever air conditioning system you install there is, in my view, a limit to the degree of control that you can have, if cost has any meaning at all. I have a case where about three years ago we moved seven hundred people out of an air conditioned building in London to the centre of Bristol away from the main traffic routes, where we took a good building which had no air conditioning although the fenestration was absolutely superb, you could say, if you like, that one could control ones destiny in such an old fashioned type of environment. A large percentage of the people who have experienced both conditions say Bristol is so much better. If they do not want fresh air they do not need to have it, and if they want a cool environment, they can get it.

Chapter 2

The Logic Evaluation and Consequence of a Total Cost Philosophy

P. E. Bathurst

The cost of a building to the building owner, although normally paid by instalments over the period of the contract, may be considered as a single capital charge. The expenditure on fuel and maintenance that will occur throughout the life of the building may be considered as annual charges. The full economic effect of the various design decisions taken by the architect can only be examined if capital and long-term costs can be represented together. For commercial developments, comparisons of capital costs of land, construction and fees must be made with the long-term running costs, maintenance and the return of income through rents. Techniques for making such calculations developed by valuation surveyors and accountants are based on the principle of compound interest. Formulae derived from this principle, although mathematically simple, have a complicated form and are difficult to use in everyday calculations. Therefore, the results are calculated for various values of interest rates and length of the building's life, and presented in tabular form. However, the arithmetic of the derivation of the tables is a useful aid to understanding their use and the development of the most frequently used tables is described below.

Compound Interest

In the simplest application, interest is calculated at the end of each year as a percentage of the amount deposited at the beginning of the year. If the interest is not withdrawn, it is added to the amount deposited and becomes the principal on which the interest is calculated in the following year.

P = principal i = rate of interest n = number of years

At the end of the first year the amount equals the principal plus the interest on the principal.

$$\text{Amount (first year)} = P + iP$$
$$= P(1+i)$$

At the end of the second year interest must be calculated on the total figure:

$$\text{Amount (second year)} = P(1+i) + iP(1+i)$$
$$= P(1+i)(1+i)$$
$$= P(1+i)^2$$

$$\text{Amount (after n years)} = P(1+i)^n$$

Amount of £1/$1

Tables of compound interest are calculated using the formula above. To calculate how a certain principal will increase over a term of years when invested at a certain rate of interest, use the Table 'Amount of £1/$1' and multiply the principal by the figure from the table appropriate to the term of years and the rate of interest.

Present Value of £1/$1

Tables of present value* are calculated using the same formula.

$$\text{Amount (after n years)} = £1/\$1 = P.(1 + i)^n$$

from which it follows that

$$P = \frac{£1/\$1}{(1+i)^n}$$

The formula for the present value therefore is the reciprocal of the formula for calculating the amount of £1. To find what principal would need to be invested to achieve a certain amount, use the Table 'Present Value of £1; the amount required is multiplied by the figure from the table appropriate to the term of years and the rate of interest. The formulae can be extended to relate a capital sum invested at the beginning of the period to annual income.

Present Value of £1/$1 Per Annum

The principal that must be invested to give a certain amount after one year, designated by $P(1)$, would be given by the equation:

$$P(1) = \frac{\text{Amount}}{(1+i)}$$

The principal necessary to provide the same amount after two years, designated by $P(2)$, would be given by:

$$P(2) = \frac{\text{Amount}}{(1+i)^2}$$

To provide equal annual sums for each of two years it would be necessary to invest:

$$P(1) + P(2) \text{ or } \frac{\text{Amount}}{(1+i)} + \frac{\text{Amount}}{(1+i)^2}$$

To provide for an annual income for each of n years it would be necessary to invest a single sum, the total principal, equal to:

$$\frac{\text{Income}}{(1+i)} + \frac{\text{Income}}{(1+i)^2} + \frac{\text{Income}}{(1+i)^{(n-1)}} + \frac{\text{Income}}{(1+i)^n}$$

This is a geometric series and the sum may be found firstly by multiplying both sides of the equation by $(1 + i)$:

$$\text{Total principal} \times (1+i) = \text{Income} + \frac{\text{Income}}{(1+i)} + \frac{\text{Income}}{(1+i)^{n-1}}$$

If the former equation is subtracted from the latter, all intermediate terms disappear with the result:

$$\text{Total principal} . (1+i) - \text{Total principal} = \text{Income} \left[1 - \frac{1}{(1+i)^n} \right]$$

$$\text{Total principal} . i = \text{Income} \frac{(1+i)^n - 1}{(1+i)^n}$$

$$\text{Total principal} = \text{Income} \frac{(1+i)^n - 1}{i(1+i)^n}$$

*The term present worth is also frequently used and is equivalent to present value.

By substituting the value of £1/$1 for the income, the formulae can be used to calculate the present value of an income of £1/$1 per annum. In this form the calculation is known as the Years Purchase (Single Rate). A table based on the formulae would show what capital sum invested at the beginning of a term of years, would be equivalent to an annual return of £1 throughout the period of a particular rate of interest. If an unlimited term of years is considered, in perpetuity, the years purchase is equivalent to 100 divided by the rate of interest.

This equation can also be used to determine the annual payments necessary to redeem a loan. In this instance, the value of £1 is substituted for the principal and the formula is rearranged to show the repayment instalments.

$$\text{Repayment (annual)} = £1 \text{ (ie the loan)} \quad \frac{i(1 + i)^n}{(1+i)^n - 1}$$

Normally, for domestic loans, repayments are required monthly and valuers will use the mortgage instalment table showing the amount necessary to redeem a loan of £100 in monthly instalments. This is based on the above equation modified as follows:

$$\text{Repayment (monthly)} = \frac{£100 . i(1+i)^n}{12 \; (1+i)^n - 1}$$

Table 1 below sets out the figures for the Present Value of £1, the Present Value of £1 per annum and its reciprocal calculated at 5% interest rate for various terms of years.

Table 1

Number of Years	Present Value of £1/$1 (Note 1)	Present Value of £1/$1 per annum (Note 2)	Reciprocal Present Value of £1/$1 per annum
1	0.952	0.952	1.050
2	0.907	1.859	0.538
3	0.864	2.723	0.367
4	0.823	3.546	0.282
5	0.784	4.329	0.231
6	0.746	5.076	0.197
7	0.711	5.786	0.173
8	0.677	6.463	0.155
9	0.645	7.108	0.141
10	0.614	7.722	0.130
11	0.585	8.306	0.120
12	0.557	8.863	0.113
13	0.530	9.394	0.106
14	0.505	9.899	0.101
15	0.481	10.380	0.096
20	0.377	12.462	0.080
25	0.295	14.094	0.071
30	0.231	15.372	0.065
35	0.181	16.374	0.061
40	0.142	17.159	0.058
45	0.111	17.774	0.056
50	0.087	18.256	0.055
55	0.068	18.633	0.054
60	0.054	18.929	0.053
80		19.596	

Note 1. The amount that must be invested now to produce £1 at the end of a period of x years.

Note 2. Years Purchase (Single Rate): the amount that must be invested now to produce £1 at the end of each of a period of x years.

EXAMPLES

The following examples show how this mathematical process may be applied to a typical *cost in use* problem.

Example 1

Life of building:	80 years
Component 'A':	initial and replacement costs £4,000 Life 20 years

Component 'B':   initial and replacement costs £6,000
                                                   Life 40 years

Rate of Interest: 5 per cent

Component 'A' will be replaced after 20, 40 and 60 years. The cost of each replacement can be related to a value at the beginning of the period. The relation is between the 'amount', the cost of replacement at the various times in the future, and the present value or the 'principal' in the formula:

$$\text{Principal} = \frac{\text{Amount}}{(1+i)^n}$$

The value of the initial cost and the replacement at the start of the period will be:

$$\text{Initial Cost} + \frac{\text{Replacement Cost}}{(1+i)^{20}} + \frac{\text{Replacement Cost}}{(1+i)^{40}} + \frac{\text{Replacement Cost}}{(1+i)^{60}}$$

or $£4{,}000 \left(1 + \frac{1}{1.05^{20}} + \frac{1}{1.05^{40}} + \frac{1}{1.05^{60}} \right)$

$= £4{,}000\ (1 + 0.377 + 0.142 + 0.054)$

$= £4{,}000 \times 1.573 = £6{,}292$

Using Table 1 the calculation may be set out as follows:

Present value of £1 at 5 per cent at		
	20 years	0.377
	40 years	0.142
	60 years	0.054
		0.573
	Initial Cost	1.000
		1.573

Present value of £4,000 = £4,000 × 1.573 = £6,292

The calculation for Component 'B' would be:

Present value of £1 at 5 per cent at 40 years	0.142
Initial Cost	1.000
	1.142

Present value of £6,000 = £6,852

The relationship between the initial and replacement costs of Components 'A' and 'B' = 1 : 1.089

The calculation in this form is only possible where the number of replacements can be computed. An alternative approach, based on annual costs, can be used. If it is assumed that the initial cost of the Components is borrowed as if it were a mortgage and repaid during the life of the Component, the annual cost of such repayments can be compared to measure the financial implications of the choices.

Example 2

Components 'A' and 'B' data as before

The calculation for Component 'A'. Using the column of the reciprocals of the present value per annum from Table 1 at 5% over a period of 20 years the 'loan' would need an annual repayment of £0.080 for each £1 borrowed.

Annual Cost: £4,000 × 0.080 = £320

For Component 'B': At 5% over 40 years the repayment would be £0.058 for each £1 borrowed.

Annual Cost: £6,000 × 0.058 = £348

Relationship between annual costs of Components 'A' and 'B' gives the same ratio of 1:1.089 as did the relationship of the present values. Since the replacement costs have been taken to be identical to the initial costs and the repayments must be paid throughout the life of the building, the ratio between the annual costs will be constant at all times.

The following examples show how initial costs and annual running costs can be considered together firstly in Example 3 by converting all future costs to their present value and in Example 4 by converting initial construction costs to their annual equivalent.

Example 3

Life of building:	80 years
Installation 'A':	initial and replacement costs £4,000
	Life 20 years
	Annual running costs £300
Installation 'B':	initial and replacement costs £6,000
	Life 40 years
	Annual running costs £250
Rate of Interest:	5 per cent

The present value of initial and replacement costs, as in Example 1 is £6,292. The present value of £1 per annum at 5 per cent over a period of 80 years = £19.596. Therefore the present value of £300 per annum = £300 × 19.596 = £5,879.

Total present value of Installation 'A' = £12,171

For Installation 'B' the present value of the initial and replacement costs was £6,852 as in Example 1.

Present value of running costs at 5 per cent over a period of 80 years. £250 × £19,596. = £4,899

Total present value: £11,751

Relationship between initial, replacement, and running costs of Installation 'A' and 'B' = 1 : 0.96.

15

Example 4

Installations 'A' and 'B' data as before.

Annual cost of Installation 'A' as Example 2	£320
Annual running costs	£300
	————
	£620

Annual cost of Installation 'B'	£348
Annual running cost	£250
	————
	£598

Relationship between initial, replacement and running costs of Installations 'A' and 'B' = 1:0.96.

The following example demonstrates that capital, maintenance and running costs, can be brought to a common basis and displayed for consideration in a form similar to a cost plan. The advantage of such a presentation as this is that the total cost implications of all design decisions are shown and can be examined in relation to quantity factors and the yardstick of area used in elemental cost planning.

Example 5

EXAMPLE 5

1.   Obtained by multiplying first column by 0.0905. See Table 2.
2.   Floor area 2,700m² divided by 27.
3.   Costs based on formulae:

$$\text{Fuel cost per installed kJ} = \frac{\text{Max. Hourly Heat Loss(kJ)} \times 1{,}800 \times 24 \times 0.05}{1{,}000 \times 38 \times 20 \times 0.8}$$

$$= \text{Max. Hourly Heat Loss (kJ)} \times £0.00355$$

38MJ per litre = Calorific value
0.8 = Overall efficiency
£0.05 = Cost of fuel per litre
20°C = Temperature difference
1800 = Degree days

Example 5 is based on a building that is assumed to have a total floor area of 2700m² and a capital cost of £449,275. The first columns show an elemental breakdown of this capital cost and the same figures represented as an annual equivalent cost calculated at an interest rate of 9 per cent. The present value of £1 per annum at this interest rate over a period of 60 years is £11.048. This means that an investment of this sum would produce an income of £1 in each year, or alternatively a loan of this figure would need to be repaid at a rate of £1 in each year.

Table 2 shows the present value per annum table for 9 per cent; in addition a table of reciprocals is given which indicates the repayment necessary on a loan of £1.

Table 2

Years	9 per cent Present Value Per Annum	Reciprocal	Less	Multiplier
2	1.759	0.5685	0.0905	0.4780
3	2.531	0.3951	0.0905	0.3046
5	3.890	0.2571	0.0905	0.1666
6	4.486	0.2229	0.0905	0.1324
10	6.418	0.1558	0.0905	0.0653
15	8.061	0.1241	0.0905	0.0336
20	9.129	0.1095	0.0905	0.0190
30	10.274	0.0973	0.0905	0.0068
60	11.048	0.0905		—

EXAMPLE 5

Element (Area)	INITIAL CAPITAL COSTS				REPAIRS AND REPLACEMENTS					FUELS AND CLEANING				TOTAL ANNUAL EQUIV.
	Initial Capital £	Annual Equiv. for 60 years at 9% (Note 1) £	Annual Equiv. (Note 2) £	per 100m² £	Value £	Period Years	Multiplier	Annual Equiv. for 60 years at 9% £	Annual Equiv. per 100m² £	Heat Losses per hour	Fuel Costs Oil (Note 3) £	Annual Cost £	Electricity and Cleaning Annual Costs per 100m² £	Total Annual Costs per 100m² £
Foundations (540)	21,250	1,923	71.22	+ 6.23 = 77.45						$3.6 \times 540 \times 20° \times 1.14 = 44,323$	157.46	5.83		83.28
Frame	52,500	4,751	175.96											175.96
Upper Floors	45,000	4,072	150.81											150.81
Roof (540)	18,750	1,697	62.85	+ 8.67 = 71.52	7,500	20	0.0190	142.5	5.27	$3.6 \times 540 \times 20° \times 1.59 = 61,819$	219.62	8.13		84.92
Stairs	10,000	905	33.51											33.51
Walls (930)	30,000	2,715	100.55	+ 18.73 = 119.28						$3.6 \times 930 \times 20° \times 1.99 = 133,250$	473.39	17.53		136.81
Windows (510)	32,500	2,941	108.92	+ 27.67 = 135.99 / 60.70	625	10	0.0653	40.81	1.51	$3.6 \times 510 \times 20° \times 5.25 = 192,780$	684.87	25.36		162.86
Partitions	21,250	1,923	71.22											71.22
Doors	10,000	905	33.51		500	5	0.1666	83.30	3.09					36.60
Wall Finishes	21,250	1,923	71.22		625	5	0.1666	104.12	3.86					75.08
Ceiling Finishes	21,250	1,923	71.22											71.22
Floor Finishes	31,250	2,828	104.74		12,500	15	0.0336	420.00	15.55					120.29
Decorations	10,000	905	33.51		6,250 / 3,750	6 / 3	0.1324 / 0.3046	827.50 / 1142.25	30.65 / 42.30					106.46
Sanitary Fittings	4,375	395	14.62		125	5	0.1666	20.83	.77					15.39
Cold Water	10,625	961	35.59		7,500	30	0.0068	51.00	1.89					37.48
Hot Water	8,750	791	29.29		7,500	30	0.0068	51.00	1.89					31.18
Heating	41,875	3,789	140.33	- 60.70 = 79.63	35,000	30	0.0068	238.00	8.81	$8.100m^3 \times 5 \times 20° \times 0.7 = 567,000$	2014.34	74.60		163.04
Electricity	31,250	2,828	104.74		31,250	30	0.0068	212.50	7.87					212.62
Lamps	5,000	452	16.74		5,000	2	0.4780	2390.00	88.51	999,172	3549.68		100	105.25
Drainage	7,500	678	25.11											25.11
Pavings	15,000	1,357	50.25		2,500	15	0.0336	84.00	3.11					53.36
Cleaning													250	250.00
TOTALS	449,375	40,668	1506.24	1506.24				5807.81	215.10			131.46		2202.81

17

If each elemental total in the first column of Example 5 is multiplied by the reciprocal of the present value per annum, the resulting figure is the annual equivalent. This annual figure relates to the total cost and it is necessary as a second step to relate this cost to the area of the building. To give a reasonable scale to the figures the most suitable unit of measurement for this type of analysis proves to be a unit of 100 square metres. Since the area of the building is 2700 square metres, the final column in this section therefore shows the calculated annual equivalents divided by 27. Certain elements have been further adjusted to transfer part of the heating costs to other elements, a point of detail that will be explained later.

The annual equivalent cost of the replacement and maintenance items is calculated by finding the amount that would be needed annually to repay a loan equal to the cost of the replacement items during the life of those components. At an interest rate of 9 per cent a componeent with a life of 20 years would require a repayment of about 11p for each £1 borrowed; this figure is the reciprocal of the present value per annum and can be read from Table 2. This method of calculating the annual equivalent costs of the replacement items allows for a payment at the start of each period. Clearly this will also include a payment at the start of the building's life, but since this cost has been included with the initial costs, it must be deducted from those that represent the maintenance costs.

This problem is solved by calculating the annual equivalent of this particular payment related to the whole life of the building and deducting this sum from the equivalents calculated in relation to the life of the components. Reference to the column of reciprocals in Table 2 (cols 3–5) shows that the annual repayment related to the whole life of the building is 9p for each £1 borrowed. For a component with a life of 20 years, therefore, the annual equivalent of the replacements alone will be 11p minus 9p, or 2p for each £1. Table 2 (col 5) therefore gives the necessary factors.

The replacement costs assumed are set out in Example 5 together with a suggested period for the life of the components. The calculated annual equivalent has then been divided by 27 to give the costs in relation to the yardstick of 100 square metres. In general, the annual equivalent costs of the replacements are considerably less than for those of the initial costs, but this does not hold true of materials with a short life. Redecorations have a value twice the original expenditure and the replacement of light fittings is five times as expensive. The key factor is the period of replacement. Example 5 shows that a repair that becomes necessary at intervals exceeding 10 to 15 years is discounted at such a high rate that it is beginning to lose significance.

The final columns in Example 5 show the running costs of the heating system and assumed costs of electricity, and cleaning services. The heat loss calculation for the building has been split to allocate the losses against the elements where the loss occurs. The calculation of the heat required for warming the air in the building is entered against the elements 'Heating installation'. It is possible by following a formula to calculate the amount of fuel that will be necessary to support the heating requirements. Since this figure is already an annual cost, there is no need to use the discount tables, but the costs must be related to the yardstick of area. The proportion of the heat losses to the useful heat can be used to divide the cost of the heating installation between the elements. This is the adjustment made in the third column of the capital cost section. The final column of Example 5 shows the total annual equivalent element by element.

Costs presented in this form could be used to demonstrate the relation between initial replacement and running costs and to assist in the choice of specifications and design details. The elemental costs can be manipulated by the use of quantity factors in the same way as simple capital costs. For example, a change in the outline of the building, or in the proportion of the windows to walls, will alter the wall to floor ratios. If the changes in these ratios are applied to the total annual equivalent figures they will automatically adjust for the changes in heat losses and any consequent changes in the heating system. The figures can also be used to investigate the effect of changes in specification. For example, the annual running cost of fuel to offset heat losses through the windows is £684, if it were assumed that this cost might be reduced to £250 by introducing a higher standard of insulation, then over the 60 year life of the building where a capital cost of £11.05 is equivalent to an annual cost of £1.0, the saving in annual cost of £434 has a capital equivalent of £4,796. Since the windows have an area of 510 square metres the extra insulation would need to be provided at a maximum cost of £9.4 per square metre.

The examples above are based on assumed values, periods of replacement and interest rates etc. and are intended to provide a simple demonstration of the mathematical techniques used for evaluation. However, they also indicate the data necessary for such exercises and provide a basis on which the sensitivity of the methods can be judged in relation to variations in the basic information.

For a full exercise as described in Example 5 the following information is shown to be necessary:

1. An estimate of the initial cost of the components.

2. An estimate of the cost of replacing or maintaining components in the future.

3. An estimate of the future costs of fuels, labour for cleaning etc.

4. An estimate for the life of the building.

5. An estimate of the periods at which maintenance or replacement will be required for individual components.

6. The interest rate.

The first factor of initial costs is perhaps the only set of figures that can be estimated with any degree of accuracy. The range of variation will be that normally encountered by quantity surveyors in comparing estimates to tenders, and in most tendering situations will not be significant.

The second factor is much more difficult to judge mainly because of the lack of definitive information on the incidence of maintenance costs. Predicting future maintenance work must take into account three separate judgments. Firstly, with components likely to fail due to ageing or wear, the need for replacement is predictable and the costs can be estimated within reasonably close limits. Similarly, where components require repainting, these costs can be estimated assuming a planned cycle of redecorations.

The second judgment relates to the evaluation of the risk of accidental damage by storm, vandalism or mishandling and cannot be predicted other than on a basis of probability. This is likely to be the most usual reason for maintenance with components such as sanitary fittings, water waste preventors, taps, drains, ironmongery and joinery fittings. It may well be that repairs of this type cannot be directly related to the initial specifications chosen and therefore to the initial cost of the items incurred when the building was constructed. As an obvious example a drain may become blocked and it makes little difference whether it is stoneware, pitchfibre or cast iron. The overriding consideration would be the ease with which the blockage might be removed and in this case, the vital comparisons would be between the cost of clearing the drain and the initial costs of providing adequate rodding eyes or manholes, in short, matters of design rather than specification.

The third aspect requiring prediction is the effect of deferring maintenance which is indicated to be necessary by the partial failure of components. In Example 5 it is assumed that all the floors will fail after 15 years and will be replaced immediately. In reality, only partial failure is to be expected and it is likely that local repairs would be carried out over a number of years before complete replacement was thought to be necessary. In practice, materials that are regarded as having a short life may often be pieced out and repaired to give a useful life continuing until finance becomes available for their replacement. From this it can be understood that estimating future costs involves a greater degree of judgment and the balancing of many more variables than necessary for the initial capital charges.

3 — The effects of inflation will increase the apparent cost of maintenance, fuel and the labour necessary for operating plant and the cleaning services. It would not be difficult to include some arbitrary factor for price increases in the basic calculations, though obviously there would be considerable difficulty in deciding what the factor should be. The problem is that, although the procedures described are related to the rate of interest, interest is not the only 'profit' made by investment in commercial enterprise. In an expanding economy the value of investments will show a capital increase, an increase which is likely to be of the same magnitude as rises in the cost of building. In an industrial building, the very processes carried out in the new building should add to the prosperity of the firm and their ability to meet higher costs in the future.

In the long term, companies and individuals can afford to maintain their premises and expand and redevelop on new sites despite the rising cost of building. This is simply because the increased cost of building is met by the greater monetary value of their turnover and the higher incomes received. If inflation in the building industry remains generally in step with inflation in other sectors of the economy, apart from the case of a building owner who cannot expect any increase in his income, it may be ignored. However, even if this was not the case 'cost in use' exercises are mainly designed for the purpose of making reasonable judgment on specifications during the design processes. The costs of alternatives are likely to be affected by inflation to the same degree and this is probably the most urgent reason for leaving the factor out.

4 — The fourth factor listed above is the estimate of the life of the building. For a building on a leasehold site, there will be a predetermined date at which it will cease to have any value to the leaseholder, and there may be other cases where the useful life of the building will be determined by some external factor, but for most buildings the structure is likely to outlast the time span of the processes for which it is designed. The determining factors in deciding when a building will be demolished are the value of the site, the cost of adaptation or replacement to meet changing standards of safety and welfare for the occupants, and other social and economic considerations. Buildings may be said to have both a 'physical' and an 'economic' life span, and are more likely to be replaced at the end of the latter period.

Clearly, the problem of estimating the future circumstances under which the building will be valued is largely guesswork, but in fact, this particular difficulty does not greatly affect the accuracy of the procedures. The reason for this is that for any interest rate, the table of present values becomes relatively constant after about 40 or 50 years. For example, from Table 1 from year 1 to year 10 there is an increase of 6.77, but in the following 10 year periods, the differences run 4.74, 2.91, 1.78, 1.09, 0.57 etc.

5 — Errors in estimating the whole life of the building are therefore not likely to be of much importance. However, the estimate of the life of the individual components is likely to be significant for short life materials as described above. This is the fifth factor listed and the problems of estimation are similar to those described in relation to arriving at the cost of carrying out the maintenance work.

Most materials used in building, provided that they are protected from the weather, will have an almost indefinite life. The only real exceptions to this are wearing surfaces such as floors, mechanical devices such as ironmongery, service installations, and possibly specific elements of the building such as the roof, which are subject to the risk of storm damage. Therefore for these components, evidence must be sought from recorded data. In the examples earlier in this paper, it has been assumed that replacement costs will be incurred at precise intervals of time, but apart from planned preventative maintenance it is unlikely that such a regular pattern would occur.

One solution to this problem is to use probability theory. If a value is assigned to the probability of an event occuring i.e. if it is certain, a probability value of 1; if it is unlikely to happen, a value zero; particular components with a maximum life of 50 years might have a possible life pattern for all such components in the building (expressed in terms of probability) as follows:–

0—10 years	0
10—20 years	0.2
20—30 years	0.4
30—40 years	0.3
40—50 years	0.1
	1.0

The present value at each period could be calculated by multiplying the figure from the tables by the value of the replacement and the value of the probability at that time. By this means it would be certain the replacement costs of all the components would have been taken into account within a period of 50 years, but would also have taken into consideration the fact that some would have needed replacement after a life of only 10 to 20 years. This procedure might appear to give greater realism to the exercise but as was pointed out in connection with Example 5 it is only those maintenance tasks that are repeated at frequent intervals that have any real significance, and therefore this is probably an unnecessary refinement.

6 — Finally of real significance is the choice of the rate of interest. Interest rates may be considered to contain three components—risk, profit and inflation. Where there is a high rate of inflation, there will, of necessity, be a high rate of interest. As the interest rate is raised, the present value of a future payment, either annual or intermittent, is reduced, and, therefore, if a high rate of inflation is assumed the higher costs will be counter-balanced by a correspondingly high interest rate. The rate of interest is also determined by the financial position of the client. It may present the rate of interest he must pay to borrow capital, or the return that he normally expects on his capital investments. A public body may use a test rate of interest established from time to time by the Treasury. However, it will usually be sensible to carry out cost-in-use calculations using a range of interest rates. This procedure will show whether the order of preference changes at some point on the range, and it can then be considered whether this critical interest rate is relevant to the circumstances of the client.

Table 3 below sets out the results of applying three rates of interest to the data used in Example 5. It shows that as the interest rate increases, the relationship between future costs and present costs becomes of less significance. If this point is overlooked it could lead to the conclusion that high maintenance costs were acceptable simply because they had little apparent present value and omitting to take account of the resources of labour and materials that will be absorbed and which will have to be available throughout the life of the building.

Table 3

Interest Rate	Annual Equivalent of Capital Costs 100m²	Annual Equivalent of Repairs 100m²	Annual Running Costs 100m²	Total Annual Costs 100m²
5%	882	244	481	1607
9%	1506	215	481	2202
13%	2164	190	481	2835
Ratio of Future Costs to Capital				
5%	1	.2766	.5453	1.8219
9%	1	.1427	.3193	1.4621
13%	1	.0878	.2222	1.3100

I think it is clear from this discussion that much of the data needed to carry out these procedures may be difficult to determine with accuracy but, it is also necessary to realise that the objective is to produce a mathematical model for the comparative evaluation of alternative designs and specifications rather than to produce a realistic appraisal of the actual incidence of future expenditure. Reference to Example 5 shows that in some years a number of very expensive maintenance items will become necessary, yet the methodology has evened out all the irregular costs to a constant annual figure. The computation of this 'average' level is needed before comparisons can be made, but it in no way reflects the real problems that might face the building owner when the work is required and the bills fall due.

A further point that should be taken into account is the disruption that might be caused within a building whilst maintenance work is carried out. This implies that Example 5 should include further columns to represent the cost of staff employed within the building and the value of their output. In this way, loss of output or any variation in the cost of staff related to overtime or redundancy whilst production processes are interrupted, could be evaluated and included with the costs of maintenance.

The study of the relation between building and maintenance costs of buildings demonstrated in Example 5 may give some answer to the third proposition mentioned at the beginning of this paper i.e. why cost limits deal only with capital charges. In the example, the annual maintenance costs amount to about £215 per 100m². If all these costs could be offset by increased capital expenditure at 9% interest rate, the extra building costs would have to be limited to $11.048 \times £215$ per 100m². This represents about 15% of the capital costs assumed in this example. However, there are many items of maintenance that cannot be completely eliminated, especially the deterioration of service installations and wearing surfaces and therefore, the amount that could be transferred is much reduced. This means that the level of transfer from 'revenue' to 'capital' is probably limited to around 5–7½%. If this extra were applied to all buildings in the public sector at present controlled by cost limits, this would be a very considerable addition to the total expenditure. The result might well be a corresponding reduction in annual building programmes.

Notwithstanding the current economic difficulties, to the politicians, savings on expenditure in years far into the future would seem to be a poor alternative to maintaining the present quantity of public building. The reasons for limiting capital expenditure on buildings are largely political exercises in resource allocation within the economic situation current at the particular time, and this factor must be taken into account in discussing the type of control to be exercised. In fact, and this point must be made with candour, limits often exist to persuade Parliament and the public that reasonably strict financial control is being imposed during design and that public buildings should not appear to be extravagant or lavish.

The administrative style of the cost limit system used in this country has in the main been a limitation on capital costs coupled with a requirement for a minimum provision for accommodation. In some instances there may also be a requirement that the building should meet some generalised performance specification. Architects and engineers have had considerable freedom to exercise design and planning choices and it is a fact that the diversity that has flowed from this embraces a wide variety of specifications and design features.

It becomes pertinent to ask why the small percentage which might be added to the capital costs necessary to offset much of the maintenance costs, cannot be contained within the present system by limiting some of this freedom of choice. If extra money were added to capital cost limits it might be thought necessary to institute a system of standard minimum specifications to give some assurance that the extra expenditure implied by a total cost limit system was properly spent. These problems are not insoluble but they add a new dimension to the administration of systems which are even now regarded as too complicated. The apparent freedom of a total cost system might well prove to be elusive.

DISCUSSION

Dr. M. F. Green (Department of Health and Social Security)

As an example to illustrate a decision model, Mr Bathurst suggested that the decision to use stainless steel cables or painted steel for the Severn Bridge could be based upon the assessment of the cost of each. The consequential effects, such as the chance of painters falling into the river, or the disruption to traffic, being considered in a subjective manner at the end. Why not include these consequential effects in the decision model? Surely it is possible to actually cost the disruption to traffic. Surely that is the idea of decision models.

Secondly concerning the final part of the paper relating to the difficulty, yet importance of linking cost to in-use performance. Surely each aspect of a building has a particular purpose. For example sanitary appliances are provided for a particular purpose, the heating system is and so are the lifts. If we could determine the probability of the in-use performance of each of these aspects of the building being inadequate, that is identify design risks, such as the boiler size being inadequate on any day or the waiting time of lift users being intolerably long then we could begin to link capital cost to in-use performance for each aspect of the building. It might be feasible to evaluate, in a consistent manner, the consequences of each of these adverse events occurring in a building, thus completing the model linking capital cost, measurable performance and the consequences of inadequacy.

P.E. Bathurst

The answer is 'Yes' in both cases. The point I was making when I instanced the example of the Severn Bridge was simply the question of interest rates. High interest rates produce a distorted pattern. I hope I made the point later that when you make decisions on design, all aspects must be considered. I was flying a warning flag that the apparent beauty of the mathematics, and the fact that you can arrive at an answer supported by several places of decimals is, in itself, misleading unless there is certainty that all factors have been taken into account. Quite certainly falling into the Severn or disrupting the traffic should be measured and taken into account; but calculations on disruption of traffic flow involve enormous issues, starting with use of fuel, loss of business time, and even including the valuation of human life due to the accidents that will probably be caused because the traffic flow is disrupted. It really does get very difficult indeed and it is necessary to draw the line somewhere, I think this point must be at the extent of your particular client's interest. If your particular client's interest is a working motorway, then certainly you have got to take the cost of disruption into account. My warning can be put quite simply, accuracy like patriotism is not enough.

Regarding the in-use efficiency of things like lifts, I take your point but again you are in the dangerous position of stopping short of measuring the real benefits that are being obtained from these pieces of machinery. Waiting time or disruption caused by the lifts not working can be measured but are you measuring the benefits you are getting when they are working? The fact that a calculation is possible is no guarantee that the calculation will give a sensible answer or is relevant to the problem. However, the purpose of a conference such as this is to create doubts and to decide what are the relevant research issues that should be investigated. The first parts of my paper make it look as if all problems are solved—all you have to do is to pick out tables, choose interest rates, and periods of time and perform calculations. The warning I would repeatedly make is, do all that and then look at the individual factors in the calculation, measure the weight of each, and come to a conclusion whether they help you to make a decision or not.

F. G. Bennie (National Westminster Bank, Ltd. Premises Division)

One point we can tend to lose sight of sometimes when dealing with calculations and statistics is that, however you are trying to evaluate cost tomorrow with cost today you are working against certain basic assumptions on inflation, rates of interest, life etc. Judgements can only be made against the best guesses. There is nothing absolute about the figures or about the tables and conclusions produced as a result.

D. M. Andrew (National Westminster Bank Ltd.)

Building operational and maintenance costs have been reported by Milbank et al.* This research was carried out on 20 air conditioned buildings and 10 ventilated big-building complexes in the London area working over about 24 cost codes, reported weekly, over a full year. The Report has been updated to about 1973; it is available in that form, and at this moment it is being updated to 1976 figures.

P. E. Bathurst

I did wonder about including the Department of the Environment Study of Crown Offices † in my paper, but the difficulty is that they are simply figures without explanations of cause and effect. In the hospital service elaborate detailed figures are kept on the cost of raising steam and other maintenance issues. The trouble is they do not tell you why differences occur or even if consistency is to be expected at all. It is simply more data and very difficult to understand. This emphasises that the difficulty with most information on maintenance is that you can never be quite sure how the expenditure has been created. Details are available on what has been spent and when it was spent, but why it was spent and what you could have done to avoid it, is rarely reported.

M. E. Burt (Building Research Station, Garston)

I would like to discuss the apparent discrepancy between the policy implied by the financial analysis and the policy that many of us would feel from experience and instinct as architects or engineers is the right one to follow. The original capital expenditure in resource costs is multiplied by a factor of about 4–6 to get the financial payment depending on whether one takes a life of 40 or 60 years. On the other hand future maintenance and other running costs are discounted back and are divided by a large factor to get their current value. This means that the ratio of capital to running costs in current financial value in this sort of analysis can be at least 10 times the ratio in real manpower and material resource costs—admittedly at different times. Indeed, if one is discounting future costs backwards at the present interest rate of 10%, any expenditure over 20

*Milbank N. O., Dowdall J. P., Slater A. *Investigation of Maintenance and Energy Costs for Services in Office Buildings*. Current Paper CP 38/71 Dec 1971, JIHVE Vol 39 Oct 1971, pp 145-154.

'The Relationship of Capital Maintenance and Running Costs—A Case Study of Two Crown Office Buildings.' Ministry of Public Building and Works, Research and Development, Directorate of Quantity Surveying Development. 1970.
† *'Costs in Use—A Study of Twenty-Four Crown Office Buildings'* Department of the Environment: Directorate of Quantity Surveying Development. Nov 1971 (issued 1972). Property Services Agency.

years or so is getting small in current value and doubt is cast even on a 60 year life, let alone on how much should be spent on running and maintaining the building during that period.

Now this seems an uncomfortable pill for many responsible practitioners. As Mr Bennie said in presenting his paper, we are glad that some of our ancestors did not adopt the short-term view, and I suspect that in 30 or 40 years time our successors will be very grateful if we do not do so. Now, there is clearly merit in both the financial analysis and in historical experience, so one asks if there are explanations which would make the two more compatible. I suggest that there are possibly at least two.

One is that in some cases improved life or lower maintenance costs can be obtained from small increases, or even no increases, in the original cost if one knows what to do and is clever about it. An example is designing for fatigue, where larger radii can reduce stress concentrations, and greatly increase life for virtually no cost. The other lies in the longer-term national implications of any general trend towards lower capital costs, shorter lives and increased maintenance and running costs. An increasing proportion of the construction industry would have to be devoted to maintenance or reconstruction and in 20 or 30 years time the resulting effects on labour and materials costs could change the assumptions on which the financial analysis was made. If the sums could be done taking the above and other practical considerations into account they might give an answer which is more in accord with experience.

P. E. Bathurst

It is very difficult to answer you. I agree with practically everything you said except your first sentence. I do not think there is a discrepancy, I have taken the capital cost and seen how much it would cost per year to borrow the money for the building. I have also found what it would cost to borrow the money for maintaining the buildings. These calculations are absolutely on all fours but I agree with you the results may not be the ones that are really useful in making decisions.

J. Ballance (Allied Irish Banks, Dublin)

I would just like to refer to one aspect of the total cost which has not been mentioned. I refer to commercial clients and the matter of taxation. Unfortunately capital outlay is not allowable against tax but a lot of the services installations and other aspects of the interior such as carpets have depreciation allowances and also the running costs are invariably allowable against income. This factor must obviously affect the assessment of alternatives and I feel that it should be borne in mind.

P. E. Bathurst

This question and that of inflation is often raised as a problem in the systematic approach. In fact, if you follow the logic all the way through, taxation does not really make all that difference. The point is, if you decide to spend less capital on your building because you have had to pay tax on your money and allow more money to be spent on maintenance because you regard that as being offset against tax, you can do your calculations and decide what are the wise choices. The question that must be asked is what do you do with the money that you did not spend on the building? If you put up a cheap building, knowing it was going to have high maintenance costs, and then invest the money saved somewhere else, you start paying tax on that. In the end, if you do all the calculations of not only how much you spent on the building, but the alternative choices of how you spend your money on other things, you will find that taxation works out much the same in the end.

M. Curtis (University of Reading)

I would like to make some general comments on the presentation of data to a client or to a design team. Mr Bathurst has given good examples of two extremes; he has demonstrated in his presentation that very good and very sound points can be illustrated with very generalised figures. In his other example he started with a sketch on the back of an envelope, went through a calculation and ended up with figures to five significant places, although telling us that they are not accurate. I would suggest that the method of presentation of quantity surveying information tends to be in that kind of detail and inhibits other members of the design team in commenting on a particular point. Similar comments can be made about all the professions and, for example, the architect produces a drawing and it has all too soon, as Mr Bathurst said, reached a position where the client or other members of the team cannot make comment because the detail is too great.

P. E. Bathurst

I accept the implied criticism. The reason the table Mr Curtis referred to as so detailed, is for the benefit of those who are interested in how it was put together and who would like all the figures. The detail makes it easier to follow the mathematics against the tables. I will freely admit it is a failure of quantity surveyors to get into too much detail too quickly, but I think that all the training that surveyors are now getting in building economics, is producing a new breed of surveyor who can talk in broad terms about building design rather than insisting on figuring in great detail. I agree that you would not put the example in front of a client in its present form, it could be presented in percentages rather than in detailed figures, which as in the paper by Messrs Swain and Hoar might give a much clearer impression of where the balances lie. However, the first half of my paper was to show that you could evaluate in detail if you wish to, the second half was an examination of the sensitivity of the procedures.

D. Hoar (Nottinghamshire County Council)

I do not believe that there is a generalization that you can make here on how you can present the figures to the client. Obviously we would hope the new breed of quantity surveyors, Mr Bathurst has referred to, can present at the earliest stage, in a fairly fundamental and simplistic way the various cost alternatives and I would hope that those would be fairly broad and understandable to clients, but then as we progress through the sketch planning and working drawings, obviously we are refining our costs all the time and we will then be presenting more precise costs to our clients.

Chapter 3

Control of Total Cost

D. Hoar and H. T. Swain

Initial Costs

At the outset it is necessary to define the components of initial cost. In a building project the principal costs are land costs, building and engineering costs, furniture and equipment and professional fees. The relative costs and significance of these components of costs will depend largely upon the site value and the building type. An analysis of the typical costs of a Nottinghamshire primary school shows that building and engineering costs represent approximately 78% of total initial costs. Building costs of town centre projects may represent only 50% (or less), of total initial cost depending on land prices. Appendix A shows the typical breakdown of initial cost for a Nottinghamshire school.

Costs in Use

Costs-in-use are the consequential costs of running a building throughout its life. These costs-in-use include finance costs, building maintenance, cleaning, rates, electricity and fuel costs. Appendix B shows the typical costs-in-use of a Nottinghamshire school at current prices. Local Authorities usually borrow money over a forty year period to fund their initial building costs. Debt charges are related to initial costs and rates of interests, whereas the other categories of costs-in-use will generally be related to the value of the pound-sterling throughout the life of the building. The incidence of debt charges in relation to other costs-in-use throughout the building life will vary dependent upon levels of inflation and interest rates.

Appendix C shows the effect of a constant annual inflation rate of 5% per annum on the percentage of each cost-in-use category. This Appendix assumes that each 'category' cost will rise in relation to the value of the pound-sterling and illustrates the significant percentage of total costs-in-use that are attributable to debt charges—the weighted average percentage that debt charges represent over the forty year period assuming a 5% per annum inflation rate is 27%.

The Balance of Initial Costs and Costs-in-Use

Appendix B shows that the cost-in-use of a typical Local Authority building is likely to be equivalent to almost 17% of the gross building cost during the first year of the building's life. Due to the effect of inflation this percentage will decline in real terms throughout the life of the building. The client will expect his design team to advise him on the balance between quality and cost in a new building. The client will expect any additional expenditure on the building to improve standards of building performance and to be cost effective.

THE CONTROL OF TOTAL COSTS

It is possible to control total costs? An objective of a design team should be to control initial costs, but is it feasible to quantify the costs flowing from a completed building throughout the life of that building? Costs-in-use are affected by changes in interest rates and the relative costs of such categories as labour and energy. Certainly the control of capital cost is important in the public sector, due to the incidence of debt charges.

Initial and Capital Cost

Although initial costs include land costs, fees and equipment, in this chapter we will deal principally with the cost of buildings and services. We will examine the role that the participants in a building project can play in the control of initial cost. The three principal participants in a building project are the client, the design team and the construction team. The success of a building project depends upon the teamwork of these participants.

The Role of the Client

The recently published report entitled *The Public Client and the Construction Industries* (known as the 'Wood Report') devotes a section to the role of the client in a building project. Wood recognised the key role of the client in a construction project and identified the client's primary responsibilities as:

(a) The nomination of an individual to co-ordinate client requirements.

(b) The provision of a clear project brief to the design team.

(c) Monitoring the progress of the design and construction teams and involvement of the client representative in any major strategic decisions that may be required during the course of the design and construction phases.

I believe the nomination of a single senior representative of the client is of great benefit to the design team and the project. The *single* client's role is to act as a focal point for the co-ordination of the sponsor department or committee. The client must determine the relative importance of space standards, quality, capital costs, costs-in-use, programming and agree upon the overall strategy for the project. The single client must resolve the brief with his department or committee colleagues and assist the design team in the development of that brief. The dialogue between the client and the design team in the formulation and development of the brief is crucial to the optimisation of the client's requirements within the budgetary constraints imposed upon him. I will refer to this dialogue between client and designer, when I discuss the role of the design team. The client should monitor the overall progress and performance of the project, and he should make any necessary strategic client decisions swiftly, within the currency of the design or construction phases. Perhaps the most important duty of the client is the formation of the brief. This is fundamental to the success of any project.

However, the client has a further role, which is equally important. He must appoint the design team. Whether or not a project will give value for money will depend largely on the effectiveness of the design team. In this country great care is generally taken on the selection of contractors. Is equal care always taken by clients in the selection of the design team?

The Role of the Design Team

(a) *Introduction.* The client can select his design team and the contractors, who are to be invited to tender for the project. The designer, once he has accepted an appointment, has no choice but to work with the client for the project. The design team can endeavour to make the client aware of the part they must play in the enterprise. The design team, when appointed, must resolve many issues within the constraints of the client's brief. The design team will include architects, quantity surveyors and engineers. The roles and relationships of the respective members of the design team are outside the scope of this paper, but I believe that the traditional divisions that can exist between members of the design team must be overcome. The client, in his brief will have defined his objectives, and may have stated in broad terms his space and quality requirements.

(b) *The Nottinghamshire County Architect's Department.* At this point I would like to describe the way the County Architect's Department at Nottinghamshire work up the brief with one of its major clients. Nottinghamshire has an average annual new building programme of approximately £9,000,000 at current prices. In addition to this new work the County Architect is responsible for the maintenance of County buildings at an average annual cost of approximately £3,000,000 at current prices. The Department is responsible for the design of schools, social service buildings, offices and police and fire stations and is also currently engaged in the development of a component housing system in consortium with two District Councils.

The Department works in multi-discipline divisions. Architects, Quantity Surveyors and Engineers work side by side. As a Local Authority must obtain Ministry approval for loan sanction purposes on most of its building projects, capital cost is always a constraint. The design team's task at the feasibility stage is to present to the client the different design options available within the cost limitations.

(c) *Consultation with the Client.* Before any design work commences on proposed schools within a programme year, the County Architect will have informal discussions with the Director of Education and his Departmental Building Officer, who has the status of a Senior Assistant Director of Education. At this meeting the achievement of the current year in terms of space standards and quality standards are reviewed, and the general standards and the consequential cost implications for the forthcoming programme are agreed in principle. The client, on the advice of the design team, must balance his demands for standards of quality and space, so that the best value for money is achieved.

(d) *Setting Space, Performance and Quality Standards.* The establishment of the precise standards requires accurate cost advice from the design team. The levels of inflation of building costs in recent years and the erratic activity levels of the industry has made the quantity surveyor's job of cost planning more difficult. The high level of activity experienced in 1973 has been followed by a sharp decline in the demand for new buildings. The effect of activity levels on builders' prices has been dramatic.

By way of illustration it is noted that the tender price index of the Building Cost Information Service rose by only 5% in the twelve months third quarter 1974 to third quarter 1975. Over the same period factor costs for building rose by approximately 23%. In Nottinghamshire we have used Serial Contracts techniques for most of the Authority's major building programme since 1958. This method of contracting has a positive advantage in cost planning, as most of the builders' prices are known at the design stage.

(e) *Serial Contracting as an Aid to Cost Planning and Control.* Serial contracting may be defined as an arrangement whereby a series of contracts are let to a single contractor. It is principally suitable for a known programme of work over a stated period of time and in a defined area and where there is a degree of standardisation in the method of construction. Where these conditions do not exist, it is unlikely that serial contracting will offer any great advantages. Serial contracting is based on a standing offer, made by a contractor in competition, to enter into a series of lump sum contracts for stated projects in accordance with the terms and conditions set out in a *Master* or notional Bill of Quantities, which is a basic project Bill that has been extended to ensure that comprehensive ranges of price can be obtained for all categories of work.

The objective in compiling the *Master* Bill of Quantities is to reduce to a minimum the number of rates which have to be negotiated subsequently. The Contractor's tender is based on the Master Bill of Quantities and on prices ruling at the date of tender. Bills of Quantities are prepared for each project in the usual way, and these are priced at the rates contained in the Master Bill of Quantities. Allowance must be made for fluctuations in cost occurring between the date of the Master Bill tender and the contract date of each project. These fluctuations are calculated as a lump sum adjustment to the contract sum. The calculation of fluctuations are based on the NEDO (National Economic Development Organisation), indices for the price adjustment formula.

The knowledge of the Serial Contractor's prices at the design stage enables the Authority to plan the quality and standards of its building programme with more certainty, as tender levels can be accurately forecast. The design team, because of their knowledge of the Contractor's prices for each element of structure, can plan the building to use the available money to optimise the solution of the client's brief. The County Architect's Department employs about a hundred architects, quantity surveyors and engineers, plus the necessary back-up administration staff, on its new capital building programme. In an organisation of this size it is necessary to formalise certain of the key operations of the design team. The procedures for the contract cost control are defined in a Departmental Cost Planning Handbook. This Handbook, together with a cost library, is maintained by the Building Economics Group, which is a small team of quantity surveyors, who specialise in cost planning.

The Building Economics Group produce cost-data based on the prices of each Serial Contractor. This data reduces significantly the time that has to be spent on cost planning by the project teams. The Building Economics Group monitor the cost control of all major projects in the Authority's building programme.

(f) *Implementation of a Policy for Cost and Quality Standards.* I have referred previously to the dialogue which must take place between the client and design team, prior to the commencement of design work, so that the strategy in terms of objectives for costs and standards can be established. I have also referred to the problems of cost planning in the uncertain market conditions that have recently prevailed in the building industry. The Department of Education and Science suspended formal cost limits for individual projects in December 1973 in favour of a system of a *Block Allocation* to each appropriate Authority. The Block Allocation being awarded to Authorities in relation to their needs.

The consequence of this fundamental departure from formal cost limits for projects imposed by a major Ministry has been the need for Authorities to control their project costs themselves. If an Authority has, say, £3,000,000 to spend on schools in a year, it must itself decide where to build the schools and to what standards of space and quality. Authorities receive guidance from the Department of Education and Science on minimum standards, but it is for the Authority to decide whether these minimum standards are acceptable. It is important for an Authority to gauge the standards that are being achieved by other Authorities with comparable building market conditions. This information is not published and, therefore, can only be obtained by contacting other Authorities. The Department of Education and Science will provide, on an informal basis, details of space standards achieved in the regions of England and Wales and national average standards that have been achieved.

In Nottinghamshire, for a variety of reasons, we have been able to build good quality schools at costs consistently below national average costs. The weighted average saving below national cost limits and national average costs since 1958 to-date on the major primary and secondary school programmes is equivalent to approximately 4%. Many of the schools built during these years have been in excess of national average space standards, as for example all 1974 Nottinghamshire primary schools were built at approximately 15% above national average space standards.

As an average annual school programme for an Authority such as Nottinghamshire may contain twenty schools, a procedure must be developed to ensure that each project is allocated a fair slice of the total cost cake. The dialogue between the County Architect and the Director of Education culminates in a preliminary statement of design and space standards for each project being issued to project architects in the summer of the year prior to the financial year during which the project must achieve a signed contract. A copy of a typical statement is attached (see Appendix D).

This statement gives the gross internal floor area of the project based on areas per pupil or a schedule of accommodation agreed with the client. The statement also gives guidance on the quality standards that can be achieved for each element. For example, the statement will give details of the type of cladding materials, floor finishes and floor to wall ratios that form the basis of this preliminary statement. The design team are informed that this preliminary statement of quality standards is for guidance only. It is for the design team to ultimately determine how the costs are allocated to elements of the building. However, design teams are expected to contain their design within the space standards contained in this preliminary statement. A preliminary cost is included in the statement, but design teams are advised that this is preliminary only and will be subject to amendment in due course upon receipt of tenders. The purpose of this preliminary statement is, therefore, twofold:

(i) To allocate monies to projects.

(ii) To enable design work to commence with some certainty that loan sanction will be available for each project.

In January, prior to the commencement of the Programme Year in April, tenders are invited for most of the projects in the Authority's building programme. Tenders are on the basis of a Master Bill of Quantities referred to previously. The rates from this Bill of Quantities are then used to price a project Bill which reflects the design criteria detailed in the preliminary statement of space and quality standards. The results of this pricing enable the final cost limits for projects to be set at the base date of the tender. An allowance for fluctuations in the cost of labour, materials and plant has to be added to the base cost to allow for anticipated increases in cost up to the programmed date of tender for each project.

The final cost policy statement for each project is then issued to the design teams. The standards for quality and space contained in the preliminary statement are not amended at this stage.

(g) *Cost Control of the Project*. The tactical cost planning and cost control of projects is essential if value for money in building is to be achieved. The client and the architect should ensure that the project quantity surveyor is involved in the earliest discussions at the feasibility stage, for it is at the feasibility stage that the quantity surveyor can make his greatest contribution to the economy of the ultimate design solution. Quantity surveyors must charge fees for cost planning, as considerable time can be expended. However, I would submit that their fees will in most cases be more than offset by helping the architect to produce a design solution that not only complies with the client's brief, but provides good value for money. It may be thought by some architects that the quantity surveyor will inhibit their design, as his only concern is to keep capital costs to the minimum without regards to the aesthetics of the design or its performance in use. I do not accept that the quantity surveyor's contribution need inhibit good design, or that his sole objective is to reduce cost. Cost planning is about obtaining good value for money within the budget available. Cost planning, unlike architecture, is a relatively new technique and it is now taught as a subject to all students of quantity surveying and a number of good text books cover the theory. I believe quantity surveyors can and do demonstrate that they have a positive role to play in the early stages of the design, and I believe their services in this field will be increasingly sought by clients and architects.

The Role of the Contractor

(a) *Economic Design—The Contractor's View*. I have referred to the control of costs and standards through the design stage without reference to the building or engineering contractor. It is the contractor who finally decides the price of a project in his tender. Traditional methods of tendering preclude the contractor from participating in the crucial design decisions affecting costs that need to be made at the sketch planning stage. A number of techniques for early contractor selection have been used with some success. However, surveys of tendering methods in the United Kingdom suggest that the traditional tendering methods have lost little ground in recent years to new contractor selection techniques. I believe the construction team should take a hard look at methods of contractor selection with a view to finding satisfactory ways and means of making an earlier appointment of the contractor.

In bidding for work contractors must take account of their present workload and anticipated future commitments in their tenders. The building industry is a highly competitive industry and contractors must relate their profit margins in tenders to their current trading positions. Contractors with full order books will look for higher profit margins from any further commitments. Contractors short of work may reduce profits to a minimum to keep their workforce employed. Market factors certainly play a large part in the contractor's tendering strategy, but there are other considerations. One other question the contractor must attempt to resolve is the operational efficiency of the design. Will the project be easy or difficult to construct? Will one tower crane suffice or will two tower cranes be needed because of the configuration of the buildings?

The building industry is probably the only major industry in this country that separates the function of design from the process of assembly or manufacture. Therefore, it is important that the design team are mindful of the construction processes that their design solution will impose upon the contractor.

(b) *Economic Design—Research into Site Assembly Processes.* At Nottinghamshire we have been able to undertake a study of the operational efficiency of designs through a research project which we term R.S.M. (Research into Site Management). Nottinghamshire initiated the CLASP (Consortium of Local Authorities Special Programme) building system in 1957 and has, with other member authorities within the consortium, supported its continuous development. In 1967 the County Architect, Henry Swain, and his Deputy at that time, Alan Meikle, set up a research unit to focus the architect's attention on the problems of site assembly. Since the formation of the CLASP consortium, there had been good collaboration with manufacturers in the design of factory-made components for the system, and much progress had been made. However, the feedback from contractors on the problems of site assembly and operational efficiency had not been quite so good. It was considered that an improvement in the efficiency of the assembly processes would most likely lead to further economies in the construction costs. It was primarily for this reason that the research group was established. The objective of this research unit was to design and itself construct a small number of buildings using the CLASP system. The value of the buildings to be erected by the unit was not to exceed seven per cent of the County Council's major building programme. The cost limits for projects were to be identical to those for similar contracts awarded to contractors on the basis of competitive tendering.

The unit has been operating continuously since 1967, and has completed twenty-five projects. All contracts accepting system prototypes have been completed on time and within the money available. The unit employs six architects, two quantity surveyors, four supporting technical and clerical staff and twenty-eight site based operatives. It is difficult to precisely quantify the benefits of this type of ongoing research but certainly the RSM unit has provided invaluable experience for architects and quantity surveyors in operational efficient methods of building. The unit has been able to contribute to the development of the system and to the improvement of site assembly techniques. We hope to continue with this research project.

THE CONTROL OF COSTS IN USE

A client, when he buys a building, will expect his design team to control the initial cost. However, once the client occupies that building, who is responsible for controlling the cost of its upkeep? In reality, a building owner cannot control the total costs in use of a building he occupies, except possibly the short term costs. Nevertheless, the design can effect the level of costs of significant categories of total costs in use. I should like to consider each category of cost in use, and discuss the contribution that the designer can make to controlling or minimising these costs.

Financing the Project
Appendix C shows the incidence of debt charges assuming the two given variables of inflation rates and interest rates. The percentage values given in Year 1 were obtained from the Nottinghamshire County Treasurer, and are factual. The values in columns for years 10 to 40 will vary dependant upon the two variable factors mentioned. Values have not to be discounted to present values. Nottinghamshire, like most County Authorities, borrows money from two major sources. Short term loans are obtained through the London money market and long term loans from the Public Works Loan Board.

Rates of interest have risen sharply during the seventies, and the Public Works Loan Board rates of interest now current at just over 12% is almost double the rates quoted by the Loan Board four years ago.

The value of 10% given in Appendix C as the rate of interest for debt charges is currently being achieved by the Authority because of the averaging effect of long term loans secured at the lower rates of interest applicable to the fifties and sixties. The recent upturn of interest rates will increase the average interest rate paid by the Authority as old loans at lower interest are repaid, and new loans are taken up. This factor will tend to increase the incidence of debt charges.

In 1975/76 it is anticipated that debt charges in Nottinghamshire buildings will account for approximately 2 % of the total County Budget.

Building Maintenance

(a) *Building Maintenance Costs.* The maintenance of buildings in the United Kingdom claims approximately 30% of the total resources of the building construction industry. An analysis of the Department of the Environment statistics on the value of output of building work in recent years demonstrates that the share of the total building output that maintenance works represent, remains fairly constant in percentage terms.

Building maintenance is a continuing cost throughout the life of a building. The decisions made by the design team on the choice of materials and components and the method of fixing and assembly of these materials and components, will affect the cost of maintenance.

The public sector is a major building owner and building maintenance costs represent a substantial charge in their resources. Nottinghamshire County Council owns buildings having a total replacement value of approximately £250 million at present costs. The Authority spends approximately £3 million per annum on maintaining these buildings. This expenditure is insignificant compared with the cost of employing County Council employees. (The current cost in Nottinghamshire of employing the teachers, policemen, firemen, social workers and other officers is approximately £100 million per annum). Apart from the County's staff, the buildings are one of the Authority's principal assets. It is important that they are maintained to a good standard, so that they do not deteriorate prematurely and decline in value.

(b) *Building Maintenance Costs Analysed.* As a preamble to the further consideration of maintenance costs, it is necessary to examine the distribution of maintenance costs in a sample of buildings. The cost analysis of building costs will vary dependent upon the sample. Appendix E gives an analysis of the costs of maintaining the Authorities Education establishments, which includes nurseries, schools and colleges.

The Appendix indicates that in the projects sampled internal and external decoration account for about a third of total maintenance costs. In new system built projects in Nottinghamshire, plastic coated windows and internal partitions and wall linings have been introduced. These components have a higher initial cost. The use of these components, which only need to be cleaned from time to time, should reduce the incidence of decoration costs in future years.

The maintenance of roofs and floors account for about 10% of total maintenance costs. The extensive use of felt covered flat roofs has increased the percentage cost of roof maintenance substantially.

A felt covered flat roof may only have a life of about ten years, compared with an approximate life of a tiled roof of fifty years. The level of maintenance costs of roofing County Properties is currently being examined by our Chief Building Surveyor who is investigating the cost effectiveness of an improved specification for flat roofs.

31

The increase in the cost of timber and the development of the various types of flexible vinyl floor coverings and low cost hard wearing carpeting, has resulted in a change from the use of wood block flooring to thin flexible floor coverings and carpets. Flexible floor tiles may have a life of fifteen to twenty years compared with an anticipated life of fifty years for a wood block type floor. Building users appreciate the colour that the modern floor coverings can bring to a space. Flexible floor coverings are easily cleaned and carpeting is comfortable and warm. Quarry tiles or wood block floors would certainly be cheaper to maintain, and would probably be cost effective in terms of maintenance costs. (Care must be taken to consider the implications of cleaning costs in this context. The relative cost of cleaning floors is certainly more significant than maintenance in most cases.)

During the last twenty years, developments in flexible floor coverings and carpeting have transformed the flooring of buildings. Economists could no doubt prove that carpeted floors are a high maintenance cost item, but if the client wants comfort, he will probably insist on carpeting, regardless of possibly higher maintenance costs.

(c) *Subsidence.* Subsidence of the ground due to mining or mineral extraction can seriously affect the stability of building structures. Under the Coal Mining (Subsidence) Act 1957 the National Coal Board are bound to make a building 'reasonably fit for the purpose for which it was being used immediately prior to the damage occurring', where damage has been caused due to mining subsidence. However, the National Coal Board is not liable under the 1957 Act for consequential losses incurred as a direct result of mining subsidence.

A number of building systems exist that will cope with the effects of mining subsidence at minimal additional initial cost. Designers of buildings on land that may in the future be affected by mineral extraction, should consider the possible losses that could be incurred due to mining operations. Mining subsidence can be expensive in terms of consequential losses and causes inconvenience to the building occupant.

(d) *Effect of Conditions of Tenure on Maintenance Costs.* It is generally supposed that people will look after items they themselves own rather better than items loaned to or hired by them. Thurley[2], referred to the research work by White in Australia on the effect of 'tenure status' on the disrepair of a large number of identical houses. Dr. White concluded from his research, that houses that were owner occupied were in a significantly better state of repair than renter-occupied dwellings. This interesting research topic reinforces the assertion that the building occupier of estate manager has a role to play in minimising maintenance costs.

The Cleaning of Buildings

Most buildings are cleaned in some way each day that they are occupied. Appendix C shows that annual cleaning costs are approximately 66% more than annual average maintenance costs. The Building Maintenance Cost Information Service through their *Occupancy Cost Analyses* give full details of occupancy costs drawn from a wide range of subscribers building types. The incidence of cleaning costs vary considerably for different building types. In most cases, however, the cost of cleaning is a substantial percentage of the total costs-in-use. How can the design team minimise the cost of cleaning the building?

The architect by his choice of floor coverings and wall finishes can endeavour to select surfaces that will be easy to clean.

The provision of adequate cleaner's sinks and cupboards will assist the cleaning staff and possibly reduce the time taken on cleaning. In large buildings, it is important that facilities for cleaning staff are dispersed throughout the building, saving time in the collection and return of cleaning materials and water.

Window cleaning costs often represent a significant percentage of total cleaning costs. The installation of track and equipment for cradles on multi-storey buildings will facilitate window cleaning and maintenance work.

Energy Costs

(a) *Introduction*. Appendices B and C show that the cost of the heating, hot water, ventilation, lighting and power for a typical school is equivalent to 13% of the costs in use listed. This cost can be compared with the total spent on building maintenance and cleaning of 16% of total costs in use.

As a nation, we are aware of the need to save energy. The Department of Energy's free booklet entitled '*Energy Saving in the Home*' contained this simple message—'It has never made sense to waste anything. But it now makes less sense than ever before to waste energy—whether it be oil, coal, gas or electricity. We simply cannot afford to throw it away any longer. Energy saving must become part of our way of life.' This country must conserve its resources of energy, but what contribution can building designers, architects and engineers make?

In the United Kingdom, approximately 46% of all energy generated is used in servicing buildings. The remaining 54% goes into industrial practices and transportation. Energy savings in buildings can, therefore, make a major contribution to the national effort.

Nottinghamshire, like many other Local Authorities and commercial enterprises, have recently appointed an Energy Conservation Officer, and set up a Working Party composed of members of all major Departments to look at energy problems as they affect our Authority. It was immediately apparent to the Working Party that some energy saving measures could be implemented immediately, whereas other proposals would be implemented in the longer term.

(b) *Short Term Measures for Energy Conservation*. I have referred previously to the possible effect of tenure status on building maintenance costs. The opportunity for saving energy in the short term lies with the building occupant. Saving energy in a given building is dependent largely on 'Good Housekeeping.' Examples of 'Good Housekeeping' energy economies are:

(i) Reduce operating temperatures for heating installations. (It has been assessed that at 18.3°C (63°F) a 0.6°C (1.0°F) saving in temperature produces a 5% saving in fuel consumption.)

(ii) Do not open windows to waste heat.

(iii) Draw window curtains on cold nights.

(iv) Switch off electric lights when possible.

(v) Boilers and other plant should be serviced on a planned basis, and well maintained.

These energy saving practices are obvious examples to building users. However, it is a problem for the major building owner to ensure that 'Good Housekeeping' in energy conservation is practised. Education of building users and the setting of standards for energy conservation will help. It is for management to see that standards are achieved, and continue to be achieved.

(c) *Medium Term Measures for Energy Conservation*. In the medium term building owners can improve their stock of existing buildings in terms of energy usage, and they can ensure that new buildings added to their stock are designed to conserve energy.

As the relative cost of fuel rises, it becomes increasingly economic to provide better insulation standards in buildings. Building owners will expect their investment in insulation to be cost effective. Additional insulation in roof spaces and cavity insulation in hollow brick walls can improve the heat losses from a building significantly, and these improvements in insulation standards are generally cost effective. Double glazing will also reduce heat losses, but due to the relatively high initial cost, it is generally regarded as not being a cost effective investment at the present time.

Recent developments in control systems for heating installations can contribute substantial savings in fuel. Most existing buildings which are heated intermittently, are controlled by either a *night set back* or a *night switch off* with frost control, and have a fixed *start* time. Both methods of control have disadvantages. The recent development in controls known as the *optimum time start controller* can produce significant savings in fuel usage. This device, when fed with data related to internal and external temperatures, takes account of the thermal inertia of both the building and the heating system, and will compute the optimum preoccupation heating period. Trial installations of the optimum time start controller in office buildings have achieved fuel savings ranging from 15% to 50% and averaging 25%. Whilst savings of this order are unlikely to be achieved in new buildings, 'optimum time start controllers' represent a valuable aid to fuel economy. The installed cost of these controls are dependent upon the size of the installation, but they are of the order of £500 to £1,000.

(d) *Longer Term Developments*. Since antiquity, man has attempted to harness naturally occuring energy for his use. Windmills are known to have been used in Persia in the seventh century, and in England, the history of water mills extends as far back as the Norman Conquest. The technological advances made during the Industrial Revolution brought more reliable methods of producing energy, and mills powered by water or wind became obsolete. As man extracts and consumes the energy resources from the earth, the remaining reserves tend to seem more expensive to extract. This trend will make solar energy and wind or water power increasingly economic.

The French have built a tidal power plant generating 240 MW at Rance in Brittany, which is operational. The power costs from this plant are currently in excess of conventional power generation costs, although the indications are that in the medium term, the cost of power generation at this tidal power plant is likely to improve in relation to conventional methods. Tidal and wind power rely on particular geographical conditions to be effective, and a recent Working Party authorised by the Council of the Institute of Fuel, which produced a report entitled *Energy for the Future*[3], concluded that tidal and wind power are unlikely to make any major impact on world energy resources, although they should, none the less, be exploited wherever local conditions allow.

Probably the most frequent suggested source of untapped energy is solar energy. Solar energy collectors do not benefit from an increase in scale, which means that solar energy is more likely to be used in small scale applications.

Perhaps in the future, solar energy will make a significant contribution to meeting the energy needs of domestic properties.

Vandalism

Vandalism is a serious and expensive problem for certain building owners. During the last eighteen months, records have been kept of the incidence of vandalism at Nottinghamshire County Council owned properties. These records show that school buildings receive the harshest treatment from vandals, and that the incidence of vandalism at Secondary and Comprehensive schools is approximately treble the cost of vandalism at Primary establishments. The school holiday periods and the winter months, with the dark evenings, increase the level of vandalism. It is also apparent from our survey that the problem of vandalism is greater in urban areas than in rural areas.

The maintenance of records of vandalism, giving details of the damage and the date and the time of day of the occurence, can be useful to major building owners. Architects and estate managers can use these statistics in the design and subsequent management of buildings.

CONCLUSION

In this paper we have attempted to describe how Total Cost can be controlled. As Local Government Officers we have drawn from our experience in the public sector. We hope that others who are engaged in the private sector will find some relevance in the procedures which we adopt in the Nottinghamshire County Architect's Department. A number of questions we have left unanswered but we hope that what we have said will stimulate some thought and discussion.

The views and opinions expressed in this paper are personal views and opinions and are not the views and opinions of the Nottinghamshire County Council.

In conclusion, we acknowledge the help received in preparing this paper from Henry Morris, the Director of Development, of the CLASP Development Group, John Collins and other members of the Architect's Department.

REFERENCES

1   *"The Public Client and the Construction Industries."* Report of the Building and Civil Engineering Economic Development Committees Joint Working Party. HMSO 1975.
2   Thurley K. *"Responsibility to Society"*. Building Cost Information Services. March 1973.
3   *"Energy for the Future"*. Institute of Fuel, London, July 1973.

APPENDIX A.
Analysis of Typical Capital Costs of a Nottinghamshire Primary School

Category of Capital Cost	Cost Distribution for a Primary School in Village Location
Land	60%
Building and Engineering Work	770%
Furniture and Equipment	40%
Fees and Other Expenses	110%
Total	100%

Source: Nottinghamshire County Council Departmental Records

APPENDIX B
Average Total Cost-in-Use of School Buildings in Nottinghamshire*

Cost Category	Cost-in-Use per £100,000 of Gross Building Capital	Cost Category as % of Total Cost-in-Use
Average Dept Charges (based on 40 year repayment and 10% interest rate)	7,600	46
Building Maintenance	1,000	6
Caretaking and cleaning of building	1,643	10
Playing Field Maintenance	1,215	7
Rates, water rates and telephone	3,063	18
Heating, hot water, ventilation, lighting, power and cooking	2,231	13
Total	16,752	100

*Note: "Sinking Fund" costs for replacement of buildings are not included. Teaching, Clerical and School Meals Staff Salaries and central administration costs are not included. Costs are at 1975 current prices.

Land costs, furniture and equipment costs and professional fees excluded from Gross Building Capital Cost.

Sources: =Nottinghamshire County Council Departmental Records

Average projected Cost-in-Use of School Buildings in Nottinghamshire over a forty year period based on a constant future inflation rate of 5% per annum and a constant interest rate on debts of 10% per annum.

Cost Category	Cost Category as percentage of total cost-in-use					Yearly average for 40 year per.
	Year 1	Year 10	Year 20	Year 30	Year 40	
Dept Charges	46	35	25	17	11	27
Building Maintenance	6	7	8	9	10	8
Caretaking and Cleaning of building	10	12	14	15	16	13
Playing Field Maintenance	7	8	10	11	12	10
Rates, water rates and telephone	18	22	25	28	30	24
Heating, hot water, ventilation, lighting, power and cooking	13	16	18	20	21	18
Total Cost-in-Use %	100	100	100	100	100	100
Index of Total Cost-in-Use Year 1 = 100	100	84	73	66	62	75

Note: "Sinking Fund" costs for replacement of buildings are not included. Teaching, Clerical and School Meals Staff salaries and central administration costs are not included.

Sources: Nottinghamshire County Council Departmental Records

Appendix D

Specimen of Outline Specification which forms part of the Preliminary Statement of Cost and Quality Standards sent to Project Design Teams at the Feasibility Stage.

Nottinghamshire County Architect's Department
Outline Specification forming basis of Cost Policy for 1975/76 Primary Schools (this outline specification has been based on the Nottsville Primary School, Nottsville).

Wall to Floor Ratio—0·74

Element B—	**Site Preparation**
	Vegetable soil 150 mm deep, excavated and deposited on site in permanent spoil heaps; ground levelled and compacted; 150 mm limestone filling levelled, blinded and compacted. Site cut and filled; maximum fall across the site 2 metres. No structural retaining walls. No limestone filling in making up levels, excavated material suitable.

Element C— **Foundations**

1. *Structure*
   Subsidence site. Slab 125 mm thick concrete reinforced with mesh reinforcement reference A252 (3·95 kg/m². Precast concrete plinth units.

2. *Floor Finishes*
   Carpet   49%
   Vinyl tile   41%
   Quarry tile   6%
   Timber raised floor   4%
   Skirtings—vinyl tile, timber and quarry.

Element D— **Frame**
Weight 17·47 kg/m²

Element E—

Roof

1. *General Areas*
   Corrugated sheet metal roofing with three layer felt roofing. Cold roof condition (not plenum heating system).
2. *Rooflights and Holes through Roof*
   34 No. 900 × 900 mm rooflight (7 No. fixed, 27 No. ventilated).
   3 No. 1200 × 1800 mm rooflights ventilated. 12 No. fans and extracts.
3. *Monopitch Rooflights*
   7 modules × 14 modules on plan combined with tank housing (an extra 7 modules × 4 modules). Steel frame included with Element D.
4. *Eaves*
   Eaves trim and bitumen felt skirting—to concrete   55%
                                                      —to timber   45%
5. *Rainwater Installation*
   14 No. outlets.
6. *Ceilings*

Minaboard	75%
Asbestolux	8%
Minatone	6%
Asbestolux plank	1%
Coffered	0%

Element F—

External Walls

1. *Cladding*
   Storey height brick panels generally 900 mm wide Brickwork to boiler house area. Timber boarding to monopitches included with monopitch rooflight.
2. *Windows*
   Storey height windows generally 1800 mm wide; 5% of window area in Oriel windows. Independent window boards. Plinth trims.
3. *Doors*
   7 No. 1800 mm wide
   4 No. 1600 mm wide
   1 No. 1200 mm wide
   Bunker boards

Cladding Area Ratios

1.	Cladding	51%
2.	Windows	41%
3.	Doors	8%

Element G—

Internal Walls

Internal wall to floor ratio—0·66

1. *Partitions*

Plasterboard	3%	
Sundeala	35%—76% of Sundeala covered with Hessian Wall Covering	
Metal	60%	
Timber	2%	

2. *Doors*
   2 No. 1600 mm wide (glazed)
   26 No. 700—900 mm wide (21 flush, 5 glazed)
3. *Screens*
   Standard glazed screens
4. *W.C. Cubicles*
   23 No. partitions with duct panels

Elements J & K—	**Furniture and Fittings**	
	Furniture to be included equivalent to £6.00 per square metre of gross internal floor area.	

Element L—	**Plumbing**

Water tanks—2 No. SC 1000
Tank housing on roof, partly connected to monopitch
W.C.s   24 No.
L.Bs.   29 No. with mirrors
Sinks   4 No. food preparation
           2 No. cleaners
           2 No. others
Shower Tray
Urinals 2 No. ranges with two urinals each. Cold Water Installation—pipework pvc. Hot Water Installation—partly electric—7 No. electric Water Heaters, pipework copper. Waste, overflow, soil and vent pipework pvc (marley)

Element M—	**Heating Installation**

Solid fuel producing hot water to feed radiators. Warm air heating system.

Element N—	**Ventilation Installation**

Extract fans to kitchen and toilet areas.

Element O—	**Electrical Installation**

Installation wired generally in PVC cables and includes functional and integrated lighting fittings, general power points, TV points, fire alarm system, water heating and equipment services.

Element R—	**Special Services**

Gas installation to kitchen equipment.

Element S—	**Drainage**

Drainage to last connecting manhole.

Note:—	The target space standards in terms of gross internal floor areas for projects, together with a plan of the project upon which the Outline Specification is based accompany this specification.

Appendix E

Analysis of Anticipated Building Maintenance Costs for Nottinghamshire Educational Establishment for Year 1975/76.

Group	*Category*	*% that category costs represent of Total Cost of Maintenance*	*% that Group Costs represent of Total Cost of Maintenance*
Maintenance of Building Structure	Maintenance of General Building structure	23·13	
	Maintenance of roofs	5·74	32·74
	Maintenance of floors	3·87	

Group	Category	% that category costs represent of Total Cost of Maintenance	% that Group Costs represent of Total Cost of Maintenance
Maintenance of External Works (Playing Field Maintenance and Garden Maintenance excluded)	Maintenance of paved areas, roads and paths.	5·07	7·80
	Maintenance of fencing and gates	2·73	
Maintenance of Services	Maintenance of heating, hot water and ventilation installation	11·65	18·14
	Maintenance of electrical installations	6·49	
Decoration	External decoration	7·29	33·88
	Internal decoration	26·59	
Sundry "Day to Day" repairs*	Sundry "Day to Day" repairs	7·44	7·44
TOTAL			100·00

*Note: Sundry "Day to Day" repairs include items that require immediate attention such as:
(a) Broken window glazing (b) Blocked drains (c) Leaking service pipes or cisterns (d) Electrical faults (e) Heating system failure (f) Gale or storm damage.

DISCUSSION

D. Fitzgerald (University of Leeds)

I have a comment concerning Appendix B, which was said to show that maintenance of school buildings is only 6% of their total cost, which is so small that it is not worth making capital expenditure to reduce it. It seems to me that the table demonstrates the very opposite. If we exclude playing field maintenance (we are concerned with buildings) and rates and telephone (which fall entirely outside the field we can influence), we find that running costs, including energy costs, are 37% of the total cost. Thus by design you can affect over a third of the cost-in-use of your school buildings, so that the table proves that design is very important in affecting running costs, and that it might very well be worth spending capital on reducing running costs.

D. Hoar

It was not my intention to give the impression that maintenance costs are not important. Obviously they are fairly significant, but in Appendix B I have separated maintenance and cleaning. We have these figures separated in our statistics. I thought it was useful to present them separately. I believe they should be considered separately but if you wish to add them together, and also add the running costs of heating—then you could have a figure of 29% for the total cost of maintenance, caretaking, heating and lighting, etc as the separate figures were available we thought it preferable to supply them.

D. M. Andrew (National Westminster Bank Ltd.)

Referring to Mr Hoar's comments on optimum start controllers and the worthwhile economies which can be made, all the large schools in London (and there are a considerable number) are being fitted with optimum start controllers, and the average cost is in the order of £1 000 per school and the saving is averaging 13% pa.— There are four different manufacturers involved in these installations.

J. J. Mackay-Lewis (Whinney, Son and Austen Hall, London)

I think a client likes to see value for money and likes to see the materials into which he has put his money. When a lot of his money is going down below the ground level in structure and services that seems to him to be not such good value for money. With modern technology it seems that buildings could be much lighter than they are at present. We have just put in 60 000 square feet of raised floor in offices areas in a Eurobank building in the City of London and we have thus saved the screed throughout the building, and that has considerably lightened the building as a result. If you look at the structure on the ground supporting a caravan and compare with the foundations of a single storey house, buildings could become much lighter and with the use of say aerated concrete and light weight cladding. Can a saving in cost be made in structure? It seems at the moment that concrete supports concrete in many modern buildings which only adds to the cost of unnecessary foundations.

K. A. Wahab (University of Ife Ife, Nigeria and University of Reading)

I would like to comment about the contribution of occupiers' participation in the process of design and cost reduction in maintenance expenditure. I am most impressed by the results quoted of Australian research, which concluded that owner occupiers are more careful in the ways they use their houses, thus achieving a lower maintenance cost. If it is true, and I agree this is what should normally happen to what one owns and cherishes, to what extent has the result of that research been applied to the Nottinghamshire County Council housing provision.

I am, of course, aware of the political implication of home ownership and public participation in the housing provision in the United Kingdom. Is it not possible for the occupiers of public accommodation to be encouraged to participate in the maintenance of their homes, by for example providing an annual rebate such that the best maintained dwellings can have their rents cut by a percentage of savings in maintenance that would have occurred had the occupiers not participated? Another way in which the participation of the occupiers could be beneficial in reducing the cost of maintenance would be to allow the tenants to run these estates, any gain from maintenance cost reduction being used to provide some other communal services to the estate itself.

Cost reduction can be made through occupiers' participation, not only in the cost-in-use and maintenance expenditure but also in the initial design. From the point of view of my country, the initial cost of building is a very delicate issue. In Nigeria we are in an almost mad rush in the housing field, and I think we should encourage occupiers' participation in order to reduce initial costs.

I would like to give two examples. Mr Bennie in his paper (pp 1–8) noted that as a floor finish, carpet was cheaper than other materials in the long run in spite of its high initial cost. There are many occasions where double flooring is used. This situation usually arises where the client has provided a fairly expensive floor finish and then the occupier decides to put another floor finish on top. For example, a carpet on terrazzo flooring, a carpet on woodblock floor etc. It seems to me that a way to reduce costs is initially to put as cheap a floor as possible and allow the occupants a choice of their own, whenever they can afford it, if they can afford it. The second example is the possible use of partial house forms whereby the house envelope is erected and the living room and kitchen as well as one bedroom and a toilet attached to the living room are finished in the usual manner, the occupiers later rendering the walls of other rooms, papering the walls and selecting the floor finish whenever they can afford it. Such a system would not only reduce the initial cost of housing but would encourage occupiers to invest more in their homes, take more care in their usage and thus save in the maintenance cost of the building.

D. Hoar

Nottinghamshire is not a housing authority so we do not have a large number of rented houses to compile these statistics about. The Australian research just reinforced the assertion that people who do own their own properties, do maintain them in a better state than people who simply rent them. The point Mr Wahab makes about self-help in capital housing projects is an interesting one. Some work has been done on this. I know there are one or two developers building houses in a partly finished state, for example, leaving rooms unplastered etc.

H.T. Swain

In a high taxation country like Sweden or Britain a case for building your own house is enormous because it is tax-free. I would say that in the quite near future we shall be in the do-it-yourself house-building business in quite a big way in this country, rather like we are in the garden centre business at the moment and for the same reason. This has already happened in some degree in Sweden and going back to what Jeremy Mackay-Lewis has said it indicates light construction because lifting 600 tons to build a semi-detached house is no ones idea of 'Do-it-yourself'.

R. J. Wilde (Westminster City Council)

The word 'feedback' has been mentioned and Mr Hoar in presenting his paper mentioned learning something from car manufacturers by the car instruction manual applied to buildings. Could I ask what success has been achieved in feeding back maintenance information to design staff both architects and building services. I am concerned with the building services and in our experience design engineers initially expressed a certain amount of resentment although this was probably due more to a feeling that listening to people who wore overalls and maintained plant would lower their status. This was finally communicated to them after a number of years by using Clerk of Works and maintenance staff meetings as a clearing house for the complaints going back to design and attempting to sustain the interest of the designer after a job is finished.

H. T. Swain

One of our problems is that my maintenance surveyors are such nice people and they think the architects are nice people too. They put right things the architects do wrongly with a kind of resigned determination and they do not actually come back and say you have made a big mistake. I think it is important to get a feeling where you say 'look you are free to criticize please do it to the benefit of us all'. Even in a single organisation like ours we do not get all the feedback I would like to have.

F. G. Bennie (National Westminster Bank Ltd., Premises Division)

We do have regular feedback from Clerks of Works right the way up through to the Senior Group Architect with regular meetings. We found initially this was very difficult to get across. Once you have it going it works and everybody sees the value of it. If you work hard at it it is invaluable for the whole team, but it is difficult to achieve.

Chapter 4

Halifax Building Society Head Office

H. W. Pearson and J. R. J. Ellis

INTRODUCTION

One aim of this book is to produce evidence to help change the majority view, and indeed the official government view, that first cost is the only financial criterion in building design. There are already a few enlightened clients for whom cost in use is the better way to judge design, and who measure cost in use including the cost to their own organisation of using a building. As the cost of owning and using a building is but a marginal one in the total salary and overheads picture of an organisation, any feature in the building which reduces staff costs and time taken to give a service or change a method of operation pays off rapidly. Our clients, the Halifax Building Society, belong to that small group of building owners who seek value for money in this wider sense.

This is a controversial building. There has been much, extremely contrasting, criticism. Sir Hugh Casson, on BBC Television's series "Spirit of the Age", said "... *the new Head Office of the Building Society ... straddles the city with splendid and completely appropriate Victorian confidence", whilst Stephen Gardiner, in the Observer, said, "At one stroke, Halifax has received a blow from which it can never recover".*

ECONOMIC DECISIONS AND THEIR EFFECT ON DESIGN

Most of the comment relates to the scale of the building in Halifax. This is the biggest building society in the world, and its rapid growth in the past decade now makes it an organisation on par with large City institutions. Facilities for deed storage and workspace were overstretched and inefficient and the need for a new head office was acute.

It might be expected that such an organisation would choose to locate its head office in the metropolitan centre and that the Society opted to stay in Halifax says a lot for the value placed on loyalty and continuity. But of relevance to this discussion is the great cost benefit gained. Staff costs remain lower for better staff in the provinces, whilst their living standards are on a par with or better than their metropolitan counterparts whose travel and housing burdens are far greater. Much evidence confirms the improvement experienced by most companies on leaving metropolitan centres in staff continuity and performance.

Staying in Halifax also meant that available funds could be devoted to providing the right building rather than spending a high proportion on site acquisition. The central Halifax site chosen was small and irregular, with consequent cost penalties in construction, but was only a fraction of the value of its metropolitan equivalent. Out of town sites in West Yorkshire were also considered, but the one chosen enables best use of public transport and the town's services, to the benefit of the staff and community as a whole.

The next most important economic decision was to design the building in such a way that it can respond to changes in management and working methods over a long period. The main purpose of the building is to handle the mortgage process—from application to completion. This involves a flow of paper requiring many departments, and a complex filing system. Large filing systems are notoriously inhibiting to spatial change of use owing to their weight and bulk. Extensive analysis of workflow and the spatial relationship between departments at the briefmaking stage convinced the Society and ourselves that the ideal workspace would be one which placed all the related departments on one level unobstructed by bulk filing. This was because business methods could be seen to be evolving rapidly for the foreseeable future, and no split of departments onto separate floors would be other than a future embarrassment.

The most compact and flexible space possible was achieved by devoting one whole floor the full size of the site (5000m²) to the general office, and nothing else. Owing to other accommodation needs at lower levels, the first level at which the whole site area is available is the third floor. The bulk filing is remotely situated in secure sub-basements, and files are delivered on electronic request, largely automatically, to a terminal in the office space. It may be seen from the section and plans that this is achieved at some cost in construction.

Also removed from the office floor are lockers and toilets, the former to the entrance hall, the latter into the structural void below the general office. The only *non productive* spaces are four corner coffee lounges, separated by the service towers from the working areas, and affording panoramic vistas of the town and surrounding country.

The third floor working area can thus accommodate up to 500 staff when the Society reaches its planned capacity limit for the building. It is uniformly well lit and serviced by underfloor power, telecommunication and data ducting with access at 1550 mm centres, allowing the introduction of all types of business machines, and free arrangement of furniture without trailing cables.

The furniture system itself is another exercise in spending somewhat more for greater advantage: the Action Office system selected enables informal grouping of people with a measure of privacy, but with complete freedom to reorganise. The cost of change is minimal.

The building has been in use now since late 1973, and the central idea of the third floor office has completely proved itself. Many large and small scale changes of layout have been effected with ease as new methods and the new possibilities illuminated by the space have been implemented. The Society expresses confidence that it can continue to develop its methods and management style without constraint.

Not all the office space is designed specifically for flexibility. A careful line was drawn between the needs of the mortgage process and the needs of executives where privacy and confidentiality have been achieved by traditional closed offices. These are placed together with the Board Room suite on the 4th floor convenient to the large office one floor below.

THE VALUE OF AMENITIES

An important benefit, difficult to translate into cash, is the provision of facilities, both functional and amenity, which encourage high productivity through job satisfaction, involvement, loyalty and enthusiasm. Abstract qualities which relate to comfort standards but which are mainly the result of subjective design quality—a blend of high environmental standards and aesthetics. Amenity facilities include a squash court, a social club with lounge, billiards, table tennis and TV rooms, and two staff restaurants. The restaurants give alternative self- or waitress service but both are open to any employee to use.

A feature of the design of the upper office floors is the use of fully glazed window walls. A conscious decision was made to take advantage of the exceptionally good views obtainable. The economical decision would have resulted in a largely solid external wall

Plate 1 *Night-time view of the new Headquarters building of the Halifax Building Society.*

Plate 2 *Main entrance*

Plate 3 *The main 5,000 sq. metre burolandschaft office situated at third floor level.*

which would have had a better performance in terms of heat loss and heat gain, but the view is one of the features most appreciated by staff. The compensatory cost in air-conditioning, double glazing, heat and glare resisting glass are considered to be justifiable.

FINISHES AND MAINTENANCE

Then there is the matter of quality of finish, both external and internal, and the way these standards relate to maintenance costs. All materials are chosen from a limited palette. Externally the solid forms and paving surfaces are clad with York Stone, the indigenous material of Halifax. The glass is bronze tinted, fixed into dark bronze anodised aluminium frames, the whole giving a monochromatic effect and the dark glazed surfaces reflecting the surrounding older buildings. The interior policy is to provide hard wearing floor and wall finishes in all areas which are constant in quality irrespective of the status of the user. Status and responsibility is reflected by privacy and a higher spatial provision, not by variable standards of finish. Carpet is used everywhere except for utility areas. The colour is the same to give unity; in areas of hard wear and where spillage may occur, carpet tile has been used. The carpet type, an 80 : 20 wool/nylon twisted pile, golden yellow in colour, is wearing exceedingly well. Wall finishes are ceramic tile for lift cores and columns, ash veneered plywood panelling in offices, and hessian cloth where colour is required. Ceilings are suspended acoustic tile, a fully demountable system to give access when necessary to services.

Standards of maintenance and 'housekeeping' are extremely high. The Society is proud of its new Head Office and makes every effort to keep it in pristine condition, a benefit both to staff and to the image of the Society as the building is much visited.

The contractual method employed was based upon competitive tenders on approximate documentation to obtain the main contractor. This was followed by a period of close co-operation and parallel working between designers and contractor during the development of the design. Thus the Society occupied the building about a year earlier than would have been normal had the traditional sequence been followed. Any disadvantages in the fine control of cost was heavily outweighed by the reduced cost of inflation.

Cost in Use to an organisation is therefore seen to be far more than just a matter of quality of construction. This matters greatly, but the building must serve by its location and design, the special needs of its users.

SERVICING—SETTING THE CONTEXT

It became obvious at an early stage in the briefing process that if the building was to respond to the needs of the Society then a high investment in engineering services would be essential. At their simplest these needs were seen as a determination to maintain a high quality of performance and service in the total sphere of business operations, a desire to provide a quality of life to the person working within the building, and thirdly for the building to have an overall design quality which will contribute to both the internal and external environments.

The Society, in attaining its present image and standing, has moved steadily along the high technology road—a trend which is unlikely to change. For example, it will come as no surprise to learn that the user's business operations have relied, heavily ,and will continue to do so, on telecommunications links for speech and data transfer, and on computers for data storage and information flow. The evolving brief for the new headquarters served to re-emphasise this trend as will be seen in many aspects of the building's design and installation hardware. Of particular significance are the extensive systems for protection and security, the air conditioning, and the large automated document handling and storage system.

At the outset, close involvement of the client was sought and the response was such that it enabled a high level of collaboration to take place and at the same time created a climate for the understanding of basic aims and decision making. Clearly the setting of engineering standards and the process of system selection is an involved and interactive process and one which is not always governed by cost equations. For example, it was decided that the occupied areas of the building would be provided with comfort standards which required the provision of air conditioning—primarily a 'quality of life' decision, although it was also recognised that its adoption to a major extent would be a pre-requisite for the deep plan office accommodation, a requirement of the brief.

By far the greatest overlay on the engineering servicing was the emphasis placed on performance, especially where this impinged directly or indirectly on the day to day operations of the Society. Performance in this context was viewed in its widest sense and related to such items as operational performance, ability of the system to respond to change, standby and back-up facilities, energy utilization, component quality, equipment life, reliability and maintainability and the like, again underlining the fact that in many instances the selection criteria for the engineering systems were headed by factors other than initial cost.

CONCEPT AND IMPACT ON THE BUILDING

Before commenting on the engineering content of the building, it might be appropriate to mention some aspects about the servicing concept adopted and its impact on the total design. From the schedule of the areas of accommodation for the whole building (contained in the Appendix), it will be noted that nearly half of the total floor area is below ground level and is taken up by accommodation for document handling and storage, car parking and plant space. Basement plant rooms house the heating and cooling plant and other major items in addition to the air treatment plants for the adjacent areas and sub-basement spaces. Ground floor and levels above are served by air treatment plants accommodated in the structural undercarriage at second floor which is also used extensively for air distribution ducting and the location of all the terminal units serving the top two floors.

Verticle movement of services is contained in the two main centre service cores and in the four staircase support towers—the only exception to this being the air supply riser ducts serving the perimeter of the fourth floor which are integrated with the window mullions of the floor below. Three levels of plant space are accommodated in the portion of the central cores where these project above roof level and are used to house cooling towers, extract ventilation plant, water storage and lift motors.

First impressions are that the exposed portions of the building facade are excessively glazed and therefore not in keeping with an IED (integrated environmental design) approach. Clearly when viewed in terms of a strict energy balance this is so, and a more efficient thermal barrier could have been employed. The chosen solution represents a calculated, balanced trade-off in return for the benefits of the panoramic views. The trade-off was quantified by determining the impact on air conditioning capital and operating costs, and comfort effect, for different glass types and shading methods, in conjunction with varying window/wall area combinations.

The study showed that the form and geometry of the building was assisting greatly in reducing the solar effect—two faces of the third and fourth floors are basically non solar as their orientation is NW and NNE, whilst the lower glazed under-building is almost completely shaded by the overhang and adjacent buildings. Furthermore, it was possible by employing variable volume to take maximum advantage in the design of the air conditioning of the fact that the solar loads for the other two faces of the upper floors, which face SE and W, are essentially non-coincident. The glass area of the upper floors varies between 85% and 75%, depending on orientation, and comprises double glazing units, the outer section being bronz tinted and the inner clear glass. Internal roller blinds are incorporated on the two solar exposed elevations.

ASPECTS OF SYSTEM DESIGN AND SELECTION

Numerous air conditioning and ventilation systems are installed and an indication of the areas served can be obtained from the information given in the Appendix. By far the largest plants are those serving the bulk of the office accommodation located in the top two floors and also at first floor level. Factors influencing the choice of system included many of the performance criteria items previously mentioned. Of significance was the desire to provide: high standards of air filtration; facilitity to serve both perimeter and core type spaces and at the same time allow transfer of surplus core heat to the perimeter using direct transfer of air; good facilities for achieving individual room and zone temperature control; an absence of extensive secondary terminal equipment in or accessible from occupied spaces; and provision for the use of outside air for cooling. The final choice was an all-air variable volume dual duct system employing hot and cold twin fan arrangements within the central plan.

Air conditioning and central cooling and heating systems are designed for recovery of a high proportion of internally generated heat, such as from lighting, occupants and equipment. Recovery methods employ both direct air transfer and utilisation of the refrigeration condenser heat. At the heart of the recovery system are two large screw type water chilling units, each with a capacity of 1284 kW. A third smaller machine is provided for low load conditions during certain unoccupied periods. Automatic thermal balance is an integral feature and ensures that optimised use is made of outside air for cooling and that heat rejection only takes place when a surplus exists; conversely, supplementary heat is added from the boiler plant only at times when the recovered heat is insufficient for the building's needs. From operating experience the balance point of

Plate 4   *This model shows the disposition of air conditioning plant and air distribution systems in the second floor plant room, and surrounding structural void.*

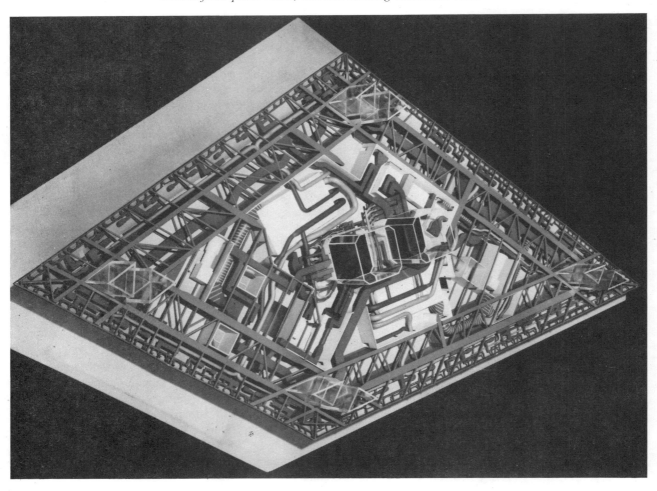

the building is about 4.5 C. The provision of heat recovery has only increased marginally the capital cost of the heating, cooling and air treatment systems and when estimated represented less than 2.5% addition to this section of cost.

The importance of telecommunications to the Society has already been stressed; it was therefore recognised that this aspect of the services should receive intensive study at an early stage to ensure that a system is selected which will provide the Society with the most appropriate level of service. A working party was set up with the client to formulate the brief and initiate traffic studies within their existing premises. The system ultimately selected is a 4000 extension, 44 exchange line crossbar PABX telephone exchange, incorporating additional facilities to benefit the Society's operation, such as flexible secretarial intercept, absent service, centralised dictation. In addition, several separate voice switched, hands free, loudspeech systems are provided to assist the business administration and operating needs of the building and include direct speech for senior management and systems for fire safety, services operation and mechanical filing operations.

In 1968 our choice of light sources was somewhat limited compared to the range presently available; furthermore, current practice at that time was saying 1000 lux for open plan offices and the lighting design adopted reflects this time datum. 1500 mm fluorescent fittings arranged in a two way grid system suspended below an acoustic ceiling provide a continuous 1000 lux nondirectional glare free light source pattern which reflects the minor planning grid. In addition, the ceiling provides acoustic performance, accommodates core supply air diffusers and also acts as a return air plenum—the extract air passing through the light fittings, for heat recovery and to reduce the amount of heat entering the space. Recent trends in the development of more efficient fluorescent light tubes are encouraging and may provide a satisfactory means of achieving lower lighting levels with commensurate savings in energy consumption.

The initial design for the building including approximately 1.0 MVA of diesel standby power plant, representing some 50% of the power loading under normal operating conditions. In late 1969 this policy was reviewed as a result of the rising unrest in the power and fuel supply industries at that time. Numerous studies were undertaken and the resulting action was to provide a third generator, thereby increasing the standby capacity to 1.5 MVA. The purpose of the added capacity was two fold—it would enable the head office to function without serious interruption in the event of a prolonged power failure, and also provide a further second standby power back up facility for the adjacent computer building. In addition, the proposals recommended increasing the oil storage, provided in the form of two underground storage tanks. It was proposed also that the natural gas burners on the boiler plant be respecified to dual fuel, therby further capitalising on the potential use of the stored oil.

The boiler plant which serves both the new head office and adjacent computer building is sized so that it can handle the total heat supply of the two buildings with reserve for standby capacity in the event of a boiler failure. Total reliance on the heat reclaim system, using the refrigeration machines for heat supply, was considered and rejected because of the possible failure risk. In addition, the provision of a 100% heat source within the boiler plant was seen as an essential back up facility which could be utilised either in the event of failure of the heat recovery machines or, more importantly, should gas or oil tariffs become more advantageous than the cost of obtaining recovery heat using electric drive on the refrigeration machines.

A whole range of special engineering support systems serve the automated document handling and storage. Extensive fire protection and detection is incorporated throughout the sub-basement area, and the storage cabinets themselves are provided with CO_2 flooding from a central storage source. In addition, the cabinets are fitted with special safety features for isolation in the event of fire and also to ensure controlled access of personnel. Power and lighting are also incorporated, together with conditioned supply and extract air for maintaining specific temperature, humidity and filtration levels which are required to ensure both safekeeping of the material stored and the efficient functioning of the electronic hardware.

To assist in the operation and maintenance of the engineering systems, a sum amounting to about 2.5% of the nett engineering cost has been invested in a computerised data centre and central control console. The centre allows for a wide variety of supervisory and management functions to be performed automatically, and accommodates controls for the environmental services, including lighting, and for fire protection and security. Facilities include monitoring, central switching, alarm indication/annunciation, automatic printout, status, visual display, etc. It is intended that the centre should also assist in the implementation of an energy conservation programme. A further sum of approximately 1% has been allocated to the purchase of a planned preventative maintenance system for the engineering installation and document handling system.

COST PROFILES

Capital

Fig 1 shows the total capital cost profile for the whole building including services. No allowance has been made in the profile for the automated document system. However, an indication of the magnitude of investment can be given by the fact that the capital cost of the system represents a sum almost equal to 30% of the total building construction amount.

Fig. 1

Total Capital Cost Profile
Cost of Construction — £.89m (£263 m²)
Equivalent Tender at Dec 75 £10.5m (£350 m²)

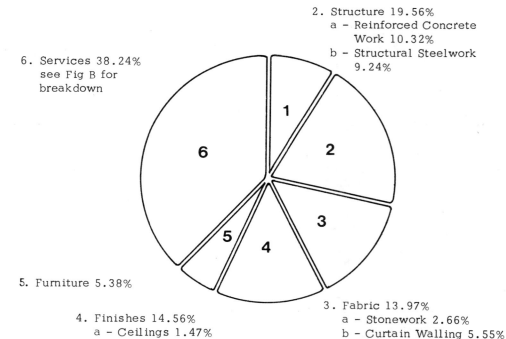

1. Substructure 8.29%
   includes Sub Basement for
   Automatic Document Handling/Storage System

2. Structure 19.56%
   a – Reinforced Concrete
       Work 10.32%
   b – Structural Steelwork
       9.24%

6. Services 38.24%
   see Fig B for
   breakdown

5. Furniture 5.38%

3. Fabric 13.97%
   a – Stonework 2.66%
   b – Curtain Walling 5.55%
   c – General 5.76%

4. Finishes 14.56%
   a – Ceilings 1.47%
   b – Joinery/Walls 10.00%
   c – Carpets 1.52%
   d – General 1.57%

Fig. 2

Engineering Services Capital Cost Profile
Total Engineering Services Cost
Cost of Installation — £3.02m
Equivalent Tender at Dec 75 — £4.71m

3. Supporting Services for Automatic Document Handling/
   Storage System 10.85%
   a – Air Treatment Systems 2.68%
   b – Electrical Services 1.43%
   c – Fire Detection and Protection CO_2 5.68%
   d – Control Systems, Data Centre and PPM 1.06%

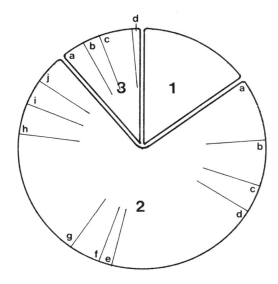

1. Services On Costs, Main
   Contractors' Profit,
   Attendances, Preliminaries,
   Builders' Work, Special
   Attendances 15.00%

2. Engineering Services (excluding Document System Supporting Services)
   74.15%
   a – Heating/Cooling Systems, including Refrig. and Boiler Plant 8.00%
   b – Water Services and Drainage 6.25%
   c – Standby Systems for Power and Boiler Fuel 2.83%
   d – Air Treatment Systems 21.00%
   e – Kitchen Equipment/Cold Rooms 1.33%
   f – Telecommunications, including Telephones, PA System, Clocks,
       Central Dictation 4.09%
   g – Electrical Installations, including Lighting 16.89%
   h – Fire Protection and Security Systems 4.11%
   i – Lift and Hoists 5.31%
   j – Control Systems, Data Centre and PPM System 4.34%

Distribution of engineering capital cost on a service by service basis is shown in Fig 2. Three main sub-sections are included to illustrate the apportionment to: the services on-costs: the nett cost of engineering services (excluding supporting services for the document system); and the nett cost of supporting services to the automated document storage and handling system.

The next diagram, Fig 3, shows again the distribution of engineering capital cost, but this time on a slightly different and perhaps more interesting basis. Each service or proportion of service has been allocated to a Category depending on the primary role which the service is provided—the definition of categories is as follows:

Category A   Basic Services providing in general comfort conditions or amenity servicing such as the heating/cooling, comfort air conditioning, water, drainage and sanitation services, lifts, basic power and lighting, kitchen equipment, control systems, etc.

Fig. 3

Engineering Capital Cost Profile

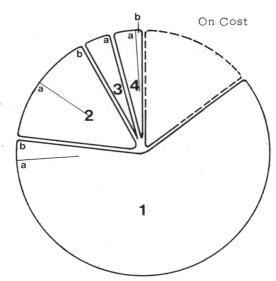

4. Category D 3.71%
   a – Building Services 2.75%
   b – Auto Document System
       Supporting Services 0.96%

3. Category C 3.62%
   a – Building Services 3.62%

2. Category B 15.06%
   a – Building Services 7.98%
   b – Auto Document System
       Supporting Services 7.08%

On Cost

1. Category A 62.61%
   a – Building Services 59.80%
   b – Auto Document System
       Supporting Services 2.81%

Category B   Protection and Security Services providing essentially protection of the building, contents, operations and occupants; includes systems such as security, fire detection and protection, standby power and boiler fuel systems, document system, air conditioning, etc.

Category C   Business Systems providing direct assistance in the business operations and activities; includes systems such as telecommunications, telephones, dictation, selective intercom, public address, etc.

Category D   Operational/Maintenance Systems to assist in the day to day operation and maintenance of the building engineering services includes data control centre and PPM system.

It is interesting to note from Fig 3 that approximately a quarter of the total nett engineering cost (22.39% out of a total 85%) is allocated to Categories B, C and D and, furthermore, that 15,06% out of a total 85% of the nett cost has been spent on Category B alone— protection and security services.

Energy, Operation and Maintenance

The building started its useful life towards the end of 1973. Since this time operation and maintenance policies have been continually developed and rationalised and the Society are now at a point where the 'learning curve' is rapidly disappearing. It will be appreciated that only when this point is passed can reliable and meaningful data be provided for feedback purposes. However, notwithstanding this comment, the Society have kindly agreed to release information gained on their operational experiences to date and these are presented in Fig 4, 5 and 6.

Fig D indicates operating costs with three main sub-divisions, namely: maintenance, cleaning and servicing; engineering and building maintenance and operating staff, including fire and security; and lastly, energy, fuel and water. The profile indicates clearly the high investment required in staffing for the four categories listed. Electricity amounts to nearly a third of the total costs whilst the high standard of office and external cleaning is achieved for about 12% of the total.

An alternative method of sub-division of operating costs is shown in Fig 5. In this diagram, maintenance and staffing have been grouped together with energy, where appropriate, for the following four sub-divisions: building engineering services: building

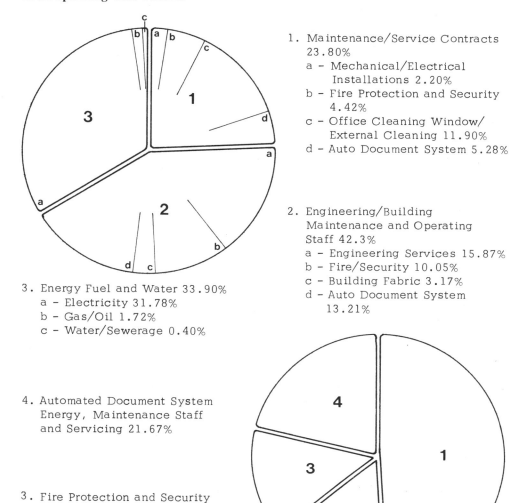

1. Maintenance/Service Contracts
   23.80%
   a – Mechanical/Electrical
        Installations 2.20%
   b – Fire Protection and Security
        4.42%
   c – Office Cleaning Window/
        External Cleaning 11.90%
   d – Auto Document System 5.28%

2. Engineering/Building
   Maintenance and Operating
   Staff 42.3%
   a – Engineering Services 15.87%
   b – Fire/Security 10.05%
   c – Building Fabric 3.17%
   d – Auto Document System
        13.21%

3. Energy Fuel and Water 33.90%
   a – Electricity 31.78%
   b – Gas/Oil 1.72%
   c – Water/Sewerage 0.40%

4. Automated Document System
   Energy, Maintenance Staff
   and Servicing 21.67%

3. Fire Protection and Security
   Staffing and Maintenance 14.47%

2. Building Fabric Maintenance and Office
   Cleaning/Window Cleaning 15.07%

1. Engineering Services,
   Energy, Maintenance and
   Operating Staff 48.79%

fabric including cleaning; fire protection and security; and automated document system. Nearly 50% of the total operating cost is expended on the energy and the operation and maintenance of the engineering services and a further 22% or thereabouts on the automated document system. The remaining percentage is split almost equally on costs associated with the building fabric and cleaning on one hand and fire protection and security on the other.

The final diagram, Fig F, shows the electrical energy profile and the sub-division between the various users. Air treatment systems are a significant proportion and reflect the continuous operation of certain plants, especially those serving the extensive basement and sub-basement areas in which the deeds and other document storage is located. Much of the pumping equipment is also in continuous use and is required essentially for serving the 24 hour operational air plants with heating and cooling energy. The operation of the refrigeration equipment on heat reclaim accounts for the higher percentage than would normally be expected with the plant operating purely on

Fig. 6   **Electrical Energy Usage Profile**

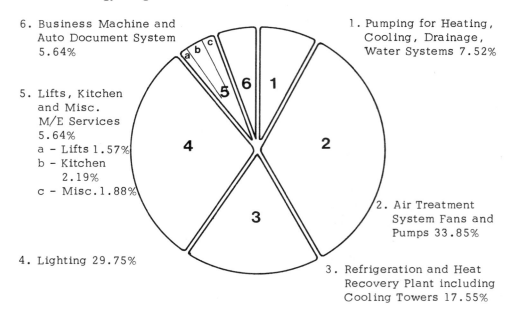

6. Business Machine and Auto Document System 5.64%

5. Lifts, Kitchen and Misc. M/E Services 5.64%
   a – Lifts 1.57%
   b – Kitchen 2.19%
   c – Misc.1.88%

4. Lighting 29.75%

1. Pumping for Heating, Cooling, Drainage, Water Systems 7.52%

2. Air Treatment System Fans and Pumps 33.85%

3. Refrigeration and Heat Recovery Plant including Cooling Towers 17.55%

cooling duty. The additional energy cost for heat reclaim operation is correspondingly offset by the reduction in primary fuel to the boilers as indicated in Fig 4. Business machinery and the automated document system account for only about 5% of the total power used. A similar amount is used by the combination of miscellaneous plant, kitchen equipment and lifts; the remaining item, lighting, represents just under a third of the overall usage.

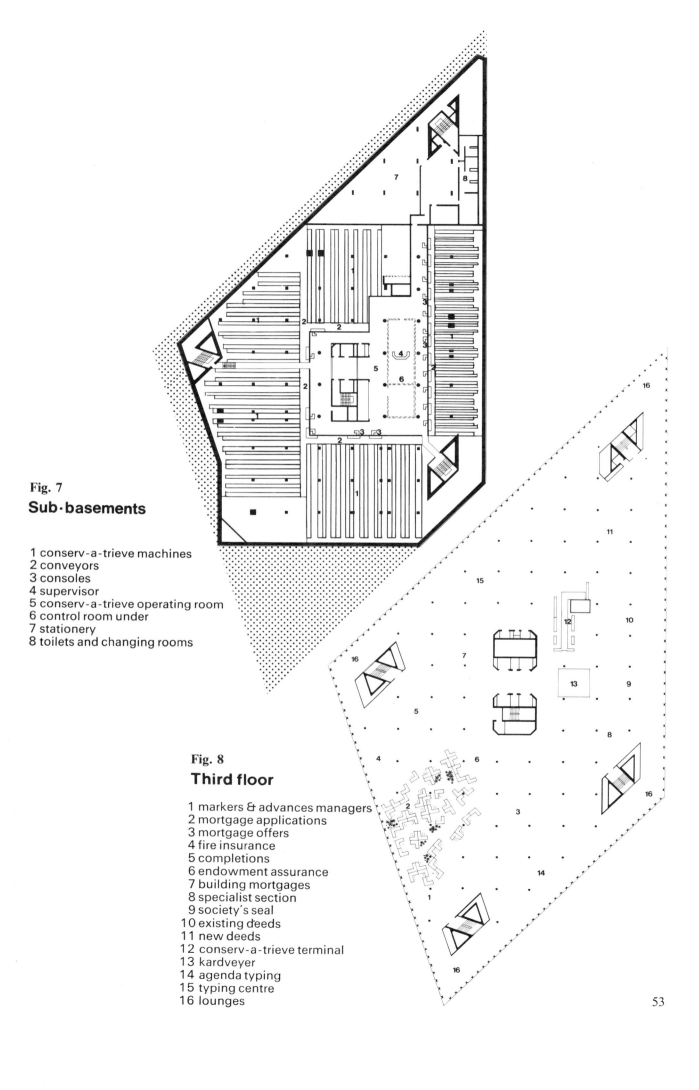

Fig. 7
Sub·basements

1 conserv-a-trieve machines
2 conveyors
3 consoles
4 supervisor
5 conserv-a-trieve operating room
6 control room under
7 stationery
8 toilets and changing rooms

Fig. 8
Third floor

1 markers & advances managers
2 mortgage applications
3 mortgage offers
4 fire insurance
5 completions
6 endowment assurance
7 building mortgages
8 specialist section
9 society's seal
10 existing deeds
11 new deeds
12 conserv-a-trieve terminal
13 kardveyer
14 agenda typing
15 typing centre
16 lounges

53

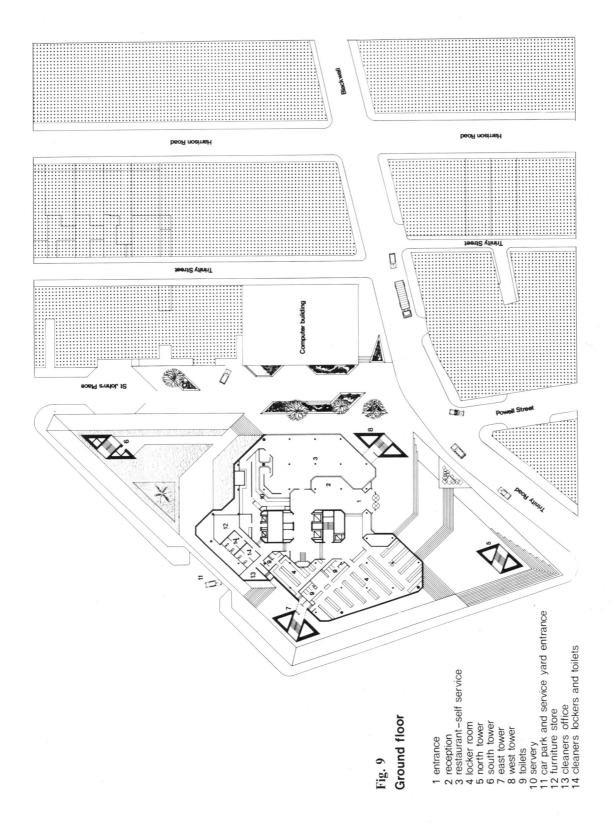

Fig. 9
Ground floor

1 entrance
2 reception
3 restaurant–self service
4 locker room
5 north tower
6 south tower
7 east tower
8 west tower
9 toilets
10 servery
11 car park and service yard entrance
12 furniture store
13 cleaners office
14 cleaners lockers and toilets

Computer building

Blackwell

Harrison Road

Harrison Road

Trinity Street

Trinity Street

St John's Place

Powell Street

Trinity Road

Fig 10

Halifax Building Society, Head Office, Long Section

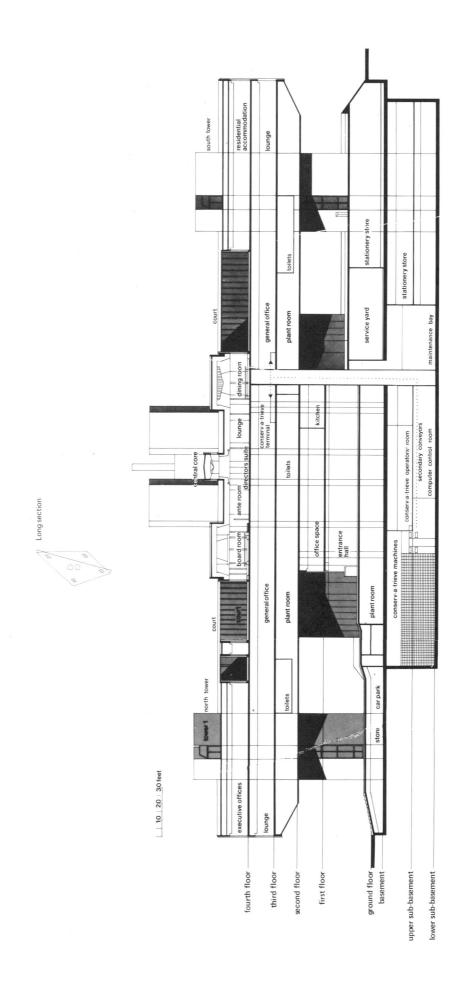

APPENDIX: SCHEDULE OF AREAS OF ACCOMMODATION

	m²	%
Areas Allocated to Business Activities		
General Office Spaces, 3 Floor	5,181	15·30
1 Floor	1,149	3·39
Executive Offices and Support Areas, 4 Floor	2,125	6·28
Directors' Suite, Conference and Board Room, 4 Floor	716	2·11
Stationery/Storage Areas, Basement	591	1·75
Upper Sub Basement	635	1·87
Lower Sub Basement	524	1·55
	10,921	32·25
Automated Document handling Storage		
Operational Floor, Upper Sub Basement	628	1·86
Storage Levels, Upper Sub Basement notional	2,745	8·11
Lower Sub Basement notional	3,262	9·63
	6,635	19·60
Amenity and Social Spaces		
Kitchen, 1 Floor	462	1·36
Restaurants and Serveries, Ground/Mezzanine Floors	837	2·47
Entrance/Reception, Ground Floor	503	1·47
Recreational Suite, Mezzanine Floor	577	1·70
Locker Rooms	665	1·97
Landscaped Courts, Roof	1,367	4·05
	4,411	13·02
Residential		
Caretaker Flats, 4 Floor	372	1·17
Directors'/Executive Overnight Accommodation	305	0·83
	677	2·00
Service Spaces—Non Plant		
Garage and Service Yard	2,251	6·64
Stores and Workshops	623	1·84
Lifts and Vertical Ducts	1,557	4·60
Staircases	714	2·12
Toilets (see note 2)		
	5,145	15·20
Plant Spaces		
Mechanical Plant Spaces	2,447	7·23
Electrical Plant Spaces	462	1·36
Document System Plant Spaces	628	1·86
	3,537	10·45
Structural Void		
Structural Void, 2 Floor, excludes usable plant space	2,533	7·48
Total Nett	25,422	78·04
Total Gross	33,859	100·00

Notes
1. *Circulation space is included in the areas stated.*
2. *Toilets, where appropriate, are included in the areas stated and are allocated on the basis of the area served. Total toilet space amounts to 710 m² (2·19%).*
3. *Net area based on usable occupied floor area excludes plant and duct space, landscape courts, and 2nd floor structural void.*

DISCUSSION

H. A. L. Ridpath (Department of the Environment)

Bad briefing, usually produces a bad solution, which is then usually an expensive solution, and therefore I believe briefing is related to the subject of this conference, and I was very interested to hear Mr Pearson explain that a great deal of time was spent on the briefing process for the Halifax building, and I wonder if he has since evaluated the resources that he put into the briefing stage in some form, for example, costs expressed as a percentage of standard fee or in some other way. I think we all realise good briefing does pay off even if we cannot quantify it, but were there any surprises or quantified cash results or returns from the extra effort you put in at the briefing stage.

H. W. Pearson

The answer to the first point is that we have not done any analysis of the time we spent. Certainly we did not charge an additional fee for doing such a comprehensive brief. It was not necessary because a very good brief does save so much time in the design process. Without a very good brief forward and backward steps seem to alternate as the client realises that he wants to make changes. It is probably more profitable to spend the time preparing a good brief, but I could not give you any analytical comment about it. As regards surprises, yes, there were some. I think mainly they were concerned with the involvement of the staff advisory group, an idea introduced by the client. This was a very good idea, incidentally, because it avoided the natural resentment often seen in staff moving into a new building, mainly because they have not been involved in the project. *'They* have done it for us'—*'they* have decided that we should be in an open office'—etc. We anticipated this and an advisory group was set up comprising all levels of staff. This group was involved in the selection of the furniture, for instance. Furniture usually causes more trouble than the design of the building itself because everyone knows about furniture and they all know what they like.

Dr D. Fitzgerald (University of Leeds)

I would like to refer to Mr Bathurst's earlier contribution. He said that the annual costs of a building, important though they be, are much less than the annual cost of the activity for which the building was built, consequently it is important to try to spend money on the building to allow the work within it to be done more effectively. The difficulty is that we do not know how to do this and, if we attempt it, it is difficult to prove that the work has been done better as a result of our extra expenditure. Bearing this in mind are there any comments on the experience of the Halifax Building Society in its new building? Have they found their output increasing? Has the money been spent in a cost-effective way as far as the organisation's purpose is concerned.

H. W. Pearson

What they say about this, and it is very difficult to measure anything to do with productivity in an office situation, is that because the building is so much more attractive to work in they are now able to attract a much higher quality of staff. They are able to be much more selective and they think that point in itself has led to greater productivity, and certainly the difference between working the paper flow in a rabbit warren up and down stairs, lots of little rooms and so on, in the old building, compared with the ease with which it flows now, is overwhelmingly more productive.

J. R. J. Ellis

A further point in addition to the apparent benefits in work output and efficiency of operation is that the client has seen a dramatic change in people's attitudes and dress— they obviously do care about the building and at the same time are making their visual contribution by looking nice.

P. J. Collins (Wylie Shanks and Partners, Glasgow)

My question concerns the energy usage. It is stated towards the end of the paper that because the refrigeration plant is working on heat reclaim there is an additional electrical energy requirement, but that this is correspondingly offset, costs-wise, by the reduction in primary fuel to the boilers. I wonder if that is totally factual and if so, are you saying that there is no real economical advantage in having a heat reclaim system— you appear to have broken even.

J. R. J. Ellis

The building is provided with various heat recovery systems which include direct recovery and transfer of air—a feature of the twin fan dual duct system employed—and also reclaim and utilisation of condenser heat from refrigeration plant. The recovery systems are fully self balancing with facilities to utilise outside air for cooling when conditions permit, thereby reducing the applied refrigeration load to a level which is sufficient to balance the heating load of the building.

We started designing eight years ago and at that time tariffs were very different to those today. The building is self sufficient in heating terms down to a balance temperature of about 4 °C; thus by using the electric drive refrigeration plant on the recovery mode we are able to satisfy a very large portion of the required heat. The cost of this heat is obviously geared to the electricity tariff, and the coefficient of performance of the plant and systems, which marginally have had the edge over alternatives such as direct gas or oil primary energy sources.

How long this situation continues is anybody's guess and it was this uncertainty, together with the policies for protection and security, which enable a number of 'option engineering' features to be incorporated. Full standby boiler plant is provided with natural gas firing so that back-up is provided in the event of heat recovery machine failure. In addition, the large bulk oil storage for the emergency generators provided a natural extension for the fitting of dual burners to the boiler plant. So all in all the client has in addition to full back-up facilities a flexibility to choose from electricity, gas or oil as the primary energy source. I think we will see more and more of 'option engineering' in buildings—all too often clients are put into an operating cost straightjacket where they cannot manoeuvre or negotiate effective tariff agreements due to inflexible options in the plant provided.

K. W. Jones (London Borough of Waltham Forest)

On the question of maintenance within the Halifax building reference has not been made in the paper to the costs for maintenance, are there any figures available? Is there a plan for protective maintenance and how much does this cost? The electrical energy costs for operating your plant seem to be high. From figures quoted you say that it is 5·64% for operating business machines and document systems, while there is 94·36% for operating the ancillary equipment, the air conditioning, air treatment, pumping etc and this includes the lifts. Lastly, have you any method of monitoring any of the faults which you find within this most complicated system?

J. R. J. Ellis

Concerning the maintenance, it might help to put Figs 4 and 5 into perspective by mentioning that they represent a total cost of some £350,000. Sector 1 of Fig 4 covers essentially maintenance and service contracts which are out of house activities—it covers items like specialised maintenance of plant and systems and cleaning which is sub-contracted out also. Sector 2 represents in house staffing for engineering and building fabric; document system; and fire and security, and approximately fifty people are involved in these activities, including management.

On the energy side, Sector 3 of Fig 4 represents the total cost of energy fuel and water. Where the energy is absorbed, is shown in the profile of Fig 6. Clearly the energy portions relating to air treatment plant and pumping appear relatively high but it must be appreciated that certain plants and equipment are operating on a 24 hour basis where this is serving the document storage spaces and adjacent building housing computer facilities.

With regard to monitoring, considerable facilities are provided. Obviously there is a learning curve to be overcome by the operational engineers but this is true of any major installation. Much of the equipment and plant incorporated is standard; however the application and hook up is somewhat unusual as are certain of the control modes. Recent trends in heat recovery systems clearly support the installation arrangements incorporated at Halifax.

E. J. Boyle (Haden Young Ltd., London)

We probably all know that maintenance, whether by the customer or professionally is rather badly done in this country and there are no accepted standards of training for maintenance engineers. That goes for the operating staff as well of course. I must add that I have never seen a more beautiful building than the Halifax both in terms of the architecture and the appearance of the plant rooms. Can you tell me who it is that you have running your plant, and what kind of men obtain the working data and analyse the figures from your monitoring system?

J. R. J. Ellis

I agree that the question of maintenance and operating documentation is important. Our policy for Halifax was a conscious one. Firstly, our contract documentation insisted that the sub-contractor provide maintenance and operating manuals and instructions, together with as installed drawings and training of the Society's engineers, all before practical completion. In the initial stages this was perfectly adequate for the needs of the operating personnel and it gave us breathing space to set up and implement the second stage and at the same time incorporate any feed back obtained during the initial operating period.

The second stage commenced some 12 months after occupation and involved the selection of a specialist consultant to provide a preventative planned maintenance system and a comprehensive set of maintenance and operating manuals and instructions for the building fabric, engineering services and document system.

We were responsible for the selection, briefing and management of the specialist and the work is almost complete. It is a considerable investment but the long term benefits are considered by the client to be worthwhile and an essential part of his day to day service operations.

The second point that you raised concerned the calibre of people responsible for the engineering services. Certain key people were engaged during the installation stages and this can be of enormous benefit as it gives them an insight into the design and visual access to services which will eventually be covered up. I might also add that our senior resident engineer was recruited into the engineering management team—it presented a wonderful opportunity to the individual and our loss was the Society's gain.

One last point, it is important to recognise at the earliest possible stage the calibre of man required to take responsibility for the engineering services and then convince the client of the need.

Professor Douglass Wise (Institute of Advanced Architectural Studies, University of York)

I think Mr Ellis is privileged in the sort of plant he has. Many of us are much less privileged. I Chaired a Conference about a month ago on energy and the point was made there, time and time again, by senior and very experienced mechanical and electrical services engineers, that the old breed of plant operator had gone, and we were in fact doing very little about replacing them. I remember one engineer telling horror stories to the conference about a building that he had been asked to go and look at recently for which he had designed the building services previously. He found that the time clocks were still set for 1971 and that 3 of the main fans had their belts missing. The question of plant operation is of paramount importance.

D. M. Andrew (National Westminster Bank Ltd.)

Referring to engineering training, this is a very big problem because we are expanding control systems and plant in size and complication. Very accurate attention is required, otherwise massive damage can result. The expansion of this kind of plant is running at about 8% a year; the availability of suitable staff is increasing at only about 4% pa. We are losing available manpower left over from the last war because most of these men are now coming close to retirement.

To deal with this problem, if you recruit a good professional engineer for the site that is big enough, part of a professional engineer's job is training, as opposed to a technician engineer's duties. A lot of Maintenance Associations have been formed round England of plant engineers and IHVE members who are interested in this problem. We find if we group apprentices or mature students and form them into classes, then the local technical colleges will train them, and they will train them to standard specifications. They will also cross-train them, to give basic electro-mechanical skills. As far as the technician-engineers are concerned, you can bring them in from contractors; you can then find time, or make time, to give them day release and to get them cross-trained. What you cannot get from contractors is information on how to deal with systems. You can find out how to deal with their individual plant very well, you can get very good handbooks, but what bothers the plant engineers running big buildings is system down time, which is either going to lose expensive production, or you can have large staffs walking out of buildings.

B. Hoskins (Mullard Ltd., Durham)

It may be of interest to you to know how we tackled this problem when setting up our Durham factory (see pp 61–82). Durham County is basically a mining and shipbuilding County and both these industries have diminished quite rapidly in recent years. Skilled men located in the area were quite obviously biased in their skills towards these industries and there is a distinct lack of the toolmaker and technician class of personnel. At Durham, we made a decision to embark on a course whereby we did not employ skilled men in specific trades.

We use their basic skills and enlarged upon these by giving training in other areas in order to convert a skilled man to a dual role craftsman. This exercise has been carried out in co-operation with the Trade Union concerned and to date has been quite successful.

The maintenance side of production is becoming increasingly more sophisticated and complex and we train both craftsmen, supervisors and production line supervisors on the machinery installed in their respective areas. To assist personnel to understand machine functions, we actively use the system developed by the Royal Navy called FIMS (Functionally Identifiable Maintenance System).

J. R. J. Ellis

A brief comment generally concerning the skill level available to operate and maintain the services installations. It is important that the designer establishes at the outset the client's attitude to engineering maintenance and operating policies and also pays careful attention to the skill levels available when determining the policy. Obviously if you are likely to experience difficulty in obtaining and/or implementing the required levels of staff then this should be borne in mind when designing the services systems—so many well engineered buildings have suffered because of inadequate operation and maintenance.

Chapter 5

Colour Television Tube Factory, County Durham, for Mullard Limited

B. J. Hoskins and J. A. Read

INTRODUCTION

The major components of total cost of an industrial development can be split into four principal areas:

(1) Research and Development

(ii) Financing

(iii) Capital Costs

(iv) Production, operating and maintenance

This case study is concerned mainly with the control of the capital costs but also deals in some part with the production, operating and maintenance costs.

Suffice it to say that the Research and Development costs can be, and often are, a significant factor in the overall cost of the product and if the financing costs are not judiciously arranged in terms of equity, loans, gearing, etc., then the economic viability of the product can be seriously affected, to the point where the product no longer becomes a commercial proposition.

Plate 1   *Aerial view of the factory*

Assuming that these first two components do not prevent viability it then follows that it is necessary to keep the capital and production costs to a minimum, commensurate with quality, in order to make the product competitive.

At the time the initial financial viability of a project is determined, the building layout, services, plant and equipment, production and operating costs are often ill defined. This is especially true in the case of a product subject to rapid technological improvements such as the manufacture of colour television tubes.

OVERALL CONCEPT

Main Process Requirements

The production of colour television tubes is a highly complex operation involving a number of separate but interlinked processes, some of which have to be carried out under stringent environmental conditions.

The process is subject to continual technological development and therefore maximum flexibility of usage of floor space within reasonably economical costs was required. This is reflected in the necessity for making modifications at any time to the conveyor routing, plant layout, services supplies and environmental controls.

The disposition of the plant, buildings and services arose principally from a consideration of the process flow requirements. Additionally the plan had to permit part of the facility to be completed at the earliest possible date and production to commence on a single line basis whilst permitting extension to three line production without interruption of the initial production line.

Early production was of such a value as to outweigh any possible additional expenditures which might be incurred by the deliberate design and phasing of construction works to meet this requirement.

Plate 2

General overhead lighting

During the period leading to the development of the Block Plan and preliminary budget estimate a cost analysis was done comparing the installation of one, two or three process lines together with the appropriate mechanical and electrical services. The sales forecasts were then examined together with the production costs and led to a decision by

Fig 1: *Mullard Limited: New Factory and Offices, Belmont, Co. Durham*

A OFFICES
A1 OFFICES
B OFFICES
C PROCESS BUILDING
D TUBEMAKING BLDG.
E SHADOW MASK BLDG.
G CANTEEN
H SERVICES BLOCK
H1 SERVICES COMPOUND
J CHEMICAL STORES
K GATE HOUSE

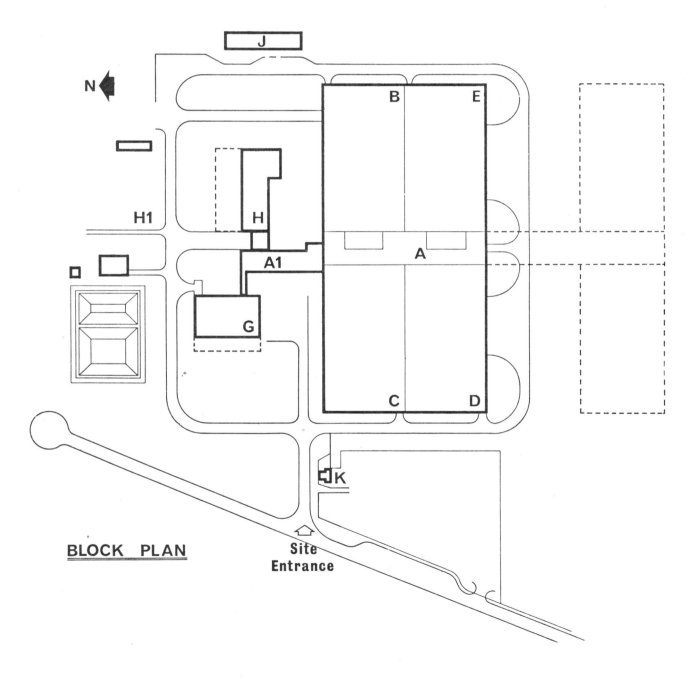

BLOCK PLAN

63

the client to install initially two flowcoating lines. A major consideration leading to this decision was the high probability of rapid technological development which could be incorporated in the third line and enable the most advanced product to be produced on this line whilst the original lines were re-organised.

The original brief was for production facilities to be divided into four main blocks with provision for extension and also the addition of blocks as required. Within the blocks the production process is served by a system of overhead conveyors which also link the blocks together. Facilities for the supply of electric power and some sixteen piped services to any point on the production floor are also incorporated.

Civil and Building

Figure 1 shows how the layout was developed, arranging the main production blocks either side of the central spine. The spine houses at ground floor level changing rooms, electrical substations, refreshment areas, cloakrooms and toilets either side of an access corridor. Factory office accommodation was provided at first floor level in this spine and the distribution of all services provided for at high level above the offices from a services block and thence out into the production buildings.

Even at an early stage of layout development it was apparent that the reduction of the outside wall areas together with the linking of production areas to office accommodation housing management and supervisory staff would lead to considerable cost savings both in terms of capital costs and also communication distances for personnel and materials. A major sociological factor was also incorporated in this plan in that by arranging the office accommodation in the centre of the production units it reduced the psychological gap between management and operating personnel.

Natural roof lighting was undesirable as some of the processes were adversely affected by certain types of light. Also it was considered preferable to keep all roof drainage external to the production buildings. Natural daylight was provided for the offices at first floor level by the provision of light wells.

The production blocks were interlinked with conveyor tunnels running through the central spine. The spine block continued to the north as an office area and houses the medical suite, main reception, conference rooms and senior management offices, etc.

Trade effluent from the processes was treated to a high degree of purity in an effluent plant before being discharged into the River Wear. Other buildings in the works complex include chemical storage buildings, pumphouses and a gatehouse, all employing steelwork in their construction for speed of erection. The construction of the main production areas was planned on a modular basis using a similar form of steelwork construction throughout, approaching a standardized building system in its simplicity.

The roof formed a support system for conveyors and accommodated air conditioning plant and ductwork, piped and electrical services, cleanroom ceilings, etc. Various patterns of the possible loadings were considered in the design and reduced to equivalent uniform loadings which formed a simple basis to which any proposed loading could be referred.

The roof was clad with galvanised steel decking insulated with a layer of expanded polystyrene weatherproofed with roofing felt, the underside being coated with a PVC organosol coating. Drainage of the roof of the production areas was achieved without the use of any separate gutters by a system of ridges at a slope of 10 degrees with valleys at 9.9 metre intervals. The whole roof falls across the width of the building at a slope of approximately 1 in 120 delivering rainwater from the valleys direct into the hopperheads of the 245 mm diameter galvanized steel drainpipes. The self-draining width is 96 metres and can readily be increased if required.

Side cladding consisted of coloured PVC plastisol coated galvanised steel sheet rolled to a square rib profile with separate internal insulation boards. Natural perimeter lighting was provided by a horizontal strip of glazing over a 3.3 metre high brick wall pierced at intervals by narrow vertical glazing strips.

The materials of the building envelope were selected not only on an initial cost basis but also to ensure low building operating costs and maintenance. Thus the roof and side cladding materials have good thermal resistivity and the finishes are such as to require the minimum of periodic painting.

Internally and externally the buildings and services were carefully designed to reduce maintenance to a minimum. All too often painting is not given sufficient attention during the design and construction stages. It tends to become an area that is considered fair game for cost slashing on the basis that maintenance costs come out of revenue. This is often a false economy as the maintenance painting costs can easily be miscalculated and do not represent a true reflection of the actual costs incurred during the life of the project. They are almost never compared back to the estimated costs.

On this project wherever possible materials were chosen that would not require painting during their operational life thereby reducing to a minimum areas to be painted. The efficacy of this policy has been more than justified: as can be seen from observation of Figure 7 maintenance painting is negligible.

Mechanical and Electrical Services

The services distribution design was an optimisation between flexibility to meet future extensions, location to give access for easy low cost maintenance and initial capital cost. The solution evolved was a high level services void which, whilst being lower in first and total cost than a partial underground system, did reflect some cost in the above ground structural element since supporting and anchoring at high level is more complex structurally than doing so in a low level duct. However, overall it was the best solution.

Plate 3   *Internal 'clean' room*

The most stringent environmental condition required for the process areas was that for the Flowcoat Clean Room Design, where the temperature limits of 21°C ± 0.5°C, humidity of 47.5% ± 2.5% and a dust count compatible with Class 10 000 rating under American Federal Standards 209 (6) was a Client requirement. Additionally it was necessary to be able to vary the air change rate between 30 and 45 air changes per hour.

The ideal technical solution to this problem would have been to use a horizontal streamlined flow clean room technique but the width requirements of the process and the need for expansion flexibility would have meant the use of large air distribution ducts at either side of each of the clean zones with its consequent high cost in additional area of building and ductwork. Since the flowcoat process required a basement to accommodate engineering services peculiar to the process itself, the possibility existed of designing the clean room on a piston flow basis of vertical air flow from ceiling to floor. This permitted a large overall space saving in accommodating the filter banks and a consequent cost saving. Filters necessary to meet the high performance requirement are expensive items to replace and a separate cost exercise was carried out on the overall benefit of pre-filtering which clearly showed that the inclusion of a bag filter (filtering to a lower efficiency) in the air flow in front of the high performance filters would be worth while. Although this increased the initial capital cost the additional amount was recovered in a very short operating period by reduction in maintenance costs.

In some aspects the quality of the services had to be necessarily high in first cost in order to satisfy process requirements, a particular case being the use of pure aluminium cooling and heating batteries in the air conditioning systems feeding the television screen production areas. It was essential that no particles of copper were present in the air flow. The use of such batteries and the use of nylon brushes on dampers, nylon pneumatic control air lines, plastic tube for de-mineralized water services and stainless steel humidifier equipment in general added to the cost of systems but this cost was of course more than out-weighed by the efficacy of the process operation which has proved to be very high.

Plate 4 *Overhead conveyors*

Specific cost exercises were carried out in the selection of lighting equipment, the final scheme adopting 2.44 m (8 ft) twin-tube flourescent fittings wherever possible. However, the tube masking area required the use of sodium lighting for process reasons and this area cost considerably more to light than elsewhere. As the security of supply to lighting in process areas was considered to be imperative, most of the lighting was designed to be supplied through static inverter equipment which cost approximately £10 000 extra over a conventional supply system.

The extent of power factor correction was established by an optimisation of the reduction in maximum demand charges against the servicing of the extra capital for automatic power factor correction equipment accomodated at each of the E.H.V. substations.

Method of Organizing Contracts

The form of contract was "1963 Private Edition (July 1969 Revision) (with quantities)" issued by the Royal Institute of British Architects. Due to the need to get the construction work started as soon as possible after the Client had decided to go ahead, the method of organizing contracts was arranged as follows:

PRELIMINARY SITE EXCAVATION–FILLING

MAIN CONTRACT

Nominated sub-contracts   Nominated Supplies
(P.C. Sums)
Steelwork
Roofing
Wall Cladding
Metal Windows and Doors
Partitions and Ceilings
Fire Resistant Doors
Monolithic Grano
Suspended Ceilings
Cross Overs
Landscaping

Mechanical Installation (Nominated specialists)
                        (P.C. and Provisional Sums)

Electrical Installation (Nominated specialists)
                        (P.C. and Provisional Sums)

This method allowed flexibility on the letting of contracts to suit the programme whilst retaining control of the specification of materials, suppliers, and costs during the project development. Bills of Quantities were progressively prepared during the early stages as the design progressed.

Appendix 1 shows the main Project Organization and Appendix 2 the main key dates.

THE COST CONTROL PLAN

General

The cost control plan was developed at the commencement of the project and was designed to enable cost variations to be highlighted continuously throughout the development, design and construction periods.

Due to the possibility of technological development throughout the design and construction period it was necessary that the cost control procedure should be developed in such a way that additions and deletions, together with changes in layout and services distribution, could be incorporated with the minimum of disruption to the plan but retaining maximum control on the highlighting of cost alterations.

Fig. 2

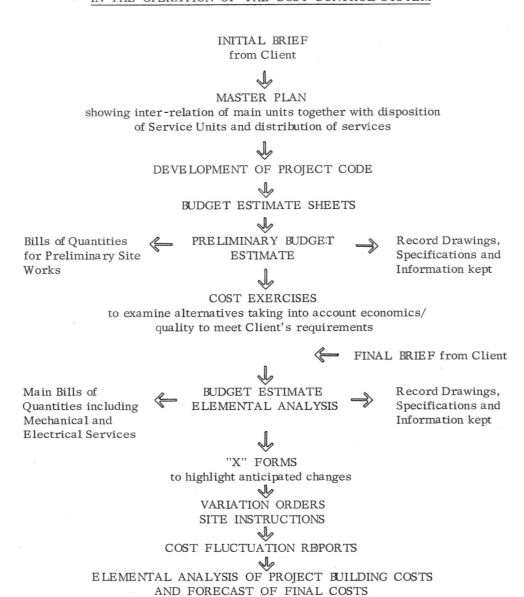

DIAGRAM SHOWING MAIN STAGES

IN THE OPERATION OF THE COST CONTROL SYSTEM

INITIAL BRIEF
from Client

⬇

MASTER PLAN
showing inter-relation of main units together with disposition
of Service Units and distribution of services

⬇

DEVELOPMENT OF PROJECT CODE

⬇

BUDGET ESTIMATE SHEETS

⬇

Bills of Quantities ⬅ PRELIMINARY BUDGET ➡ Record Drawings,
for Preliminary Site   ESTIMATE   Specifications and
Works   Information kept

⬇

COST EXERCISES
to examine alternatives taking into account economics/
quality to meet Client's requirements

⬅ FINAL BRIEF from Client

⬇

Main Bills of   ⬅ BUDGET ESTIMATE ➡ Record Drawings,
Quantities including   ELEMENTAL ANALYSIS   Specifications and
Mechanical and   Information kept
Electrical Services

⬇

"X" FORMS
to highlight anticipated changes

⬇

VARIATION ORDERS
SITE INSTRUCTIONS

⬇

COST FLUCTUATION REPORTS

⬇

ELEMENTAL ANALYSIS OF PROJECT BUILDING COSTS
AND FORECAST OF FINAL COSTS

Figure 2 shows the main stages in the operation of the cost control system.

Project Cost Code

The project was broken down and codified into sections, sub-sections and elements taking into account the requirements for programming, translation to the Philips Standard Cost Code and for abstraction of statistics by Mullards for development grant and corporation tax purposes.

There was a total of 250 separate elements in the main sections for the building services alone, which gives some indication of the level to which control can be exercised.

The main sections for the building elements were:

20 00 0	Main Production Building
40 00 0	Canteen and Link Block
50 00 0	Service Areas and Compound
60 00 0	General Site Works

These main sections were then further divided into sub-sections as follows:

21 00 0	Sub-structure
22 00 0	Superstructure
23 00 0	Internal Finishings
24 00 0	Fittings and Furnishings
25 00 0	Painting
26 00 0	Services-Plumbing and Waste Disposal
27 00 0	Services—Mechanical
28 00 0	Services—Electrical

These sub sections were further split down into their elements and an example of a total sub section for the main production buildings is as follows:

WS Atkins Code	Philips Code	
28 00 0	( )	*Services–Electrical*
28 01 0		Electrical Installations:–
28 01 1	(42.0)	H.T. Installation and Switchgear
28 01 2	(42.1)	Transformers
28 01 3	(42.2)	L.T. Distribution
28 01 4	(40.1)	Electric Lighting
28 01 5	(40.0)	Electric Lighting Fittings
28 01 6	(40.2)	Emergency Lighting
28 02 0	(45.0)	Lightning Protection
28 03 0		Communication Installations:–
28 03 1	(43.0)	Telephone Wiring and Distribution Boards
28 03 2	(44.0)	Fire Detection
28 03 3	(43.1)	Clocks, Bells, etc.
28 03 4	(43.2)	P.A. System and Intercom
28 04 0	(30.6)	H & V Supplies
28 05 0	(46.1)	Builders work in connection with electrical services
28 06 0	(46.9)	Builders profit and attendance on electrical services
28 07 0	(46.1)	Builders work in connection with plant and equipment

This detailed method of breakdown enabled costs to be closely controlled and monitored at all stages of the project after the formulation and agreement of the "budget estimate element analysis".

All estimates, Bills of Quantities, interim valuations, final accounts and programmes were related to the Project Code. The project budget was broken down into cost targets under the appropriate project code sections and costs were monitored against the targets. It was decided that Bills of Quantities would be prepared for the Mechanical and Electrical services in an effort to facilitate the control of these significant areas of work.

On this project the mechanical and electrical services constituted some 38% of the preliminary budget estimate and by the time the final account was settled represented 39% of the building and services costs.

Whilst the use of Bills of Quantities for mechanical and electrical services was not new, at the time it was more usual to treat these items as prime cost sums in the main bill.

Cost Monitoring: Control in Design Offices

Departments of the Consultants were provided with the Budget Estimate Elemental Analysis and were made responsible for minimising the capital cost of the sections of work within their control. They were made responsible to notify the Project Manager immediately any major additional cost was foreseen during the course of the development of the design.

The "X" form was developed (see Figure 3) and it was the department's responsibility to submit to the Project Manager this form complete with details and estimated cost effects of any necessary design alterations arising from:

(i) Other Departments

(ii) Mullard Limited

(iii) Outside Authorities

Fig. 3

W. S. ATKINS & PARTNERS
Woodcote Grove
Ashley Road
Epsom, Surrey

Dept. Serial No. 2936 ....
Project No. 2936 ....
Code Code Ref: ....
Date: ....

FORM "X"

MULLARD LIMITED

New Factory and Offices, Belmont, Co. Durham

To: ....
....
....
....

	Sig.	Date
Appr. by Dept. Head		
Rec'd by Proj. Man.		
Transmitted to Mull.		
Rec'd by Proj. Man.		
Appr. by Proj. Man.		

The following work requires authorisation to proceed:-

A. Source of Information, Request or Instruction

(1) Letter

(2) Drawing

(3) Other Source

B. Brief Description of Work Involved

C. Estimate of Costs Incurred

(1) Capital Costs Code Cost

(2) Fee Costs

D. Increase/Decrease against Capital Budget

signed ....

cc: Thornton-Firkin, Chesterfield
Thornton-Firkin, Belmont Site

The "X" form was completed immediately the change was foreseen, even if the cost could not be properly assessed at that time, and copies were distributed to the Quantity

Surveyors (to provide estimates where required), the Client and the Project Manager for approval. This was particularly relevant where alternative design comparisons were under consideration.

The Quantity Surveyors examined all construction drawings issued to site and were responsible for reporting immediately any deviation from the budget design.

Cost Monitoring: Control on Site

The Engineer's instructions to the contractor on site were confined to the following forms:

(i) Approved drawings marked and stamped "for construction."

(ii) Variation orders from the Engineer

(iii) Site instructions

(iv) Letters direct from the Engineer (limited usage)

The Project Manager delegated certain powers of authority to the Resident Engineer to order variations. This power was restricted to the use of site instructions only. It was made clear, in writing, that the contents of the letters or Minutes did not constitute an order entitling the contractor to any additional payment and it was the contractor's responsibility to request confirmation by site instruction of any order given by letter or verbally which he considered to constitute a variation.

Site instructions included the cost code reference quoted to the lowest level possible and all copies, other than the contractor's, contained an estimate of the addition or reduction in cost of the contract. This estimate of cost was prepared by the Resident Engineer in association with the Quantity Surveyor's site staff.

All site instructions were numbered consecutively and if a form was defaced the copies were endorsed, cancelled and distributed to all parties.

Although no limit was placed on the value of work for a variation on a site instruction the authority given to the Resident Engineer by the Project Manager, without reference back to the Project Manager or Project Engineer, was limited to an estimated value of £250.

Site instructions were valued by the Quantity Surveyor and agreed with the contractor each month for inclusion in the Quantity Surveyor's valuations. Each site instruction was itemised separately both in the valuation and in the Quantity Surveyor's monthly report to the Client and the Consultant.

Cost Reporting

Once the budget estimate element analysis was agreed a report on capital cost was submitted each month to the Client and these were numbered consecutively.

This capital cost report took the form of two documents:

(i) Cost Fluctuation Report

(ii) Elemental analysis of project building costs and forecast of final costs.

Fig. 4

MULLARD LIMITED

Belmont, Co. Durham

COST FLUCTUATION REPORT NO........ Date ....

WORKS SECTION .... Folio ....

NO.	CODE	HISTORY				ESTIMATED VALUES				REMARKS
		V.O.'s	S.I.'s	Re-appraised Items	Anticipated Changes	INDIVIDUAL		CUMULATIVE		
						-	+	-	+	
						£	£	£	£	

71

The cost fluctuation report showed the history of the cost fluctuation together with its estimated plus or minus value and also showed the running cumulative total. An example of the format of this cost fluctuation report is shown as Figure 4. A summary sheet was always included which showed the project building cost from the agreed budget estimate element analysis together with savings and additions to date and an estimated final cost (See Figure 5.).

The elemental analysis of project building costs and forecast of final costs was laid out in the same form as the budget estimate element analysis but included the plus or minus fluctuations of each individual item at that date, together with its estimated final cost.

It was therefore possible throughout the period of the project to compare the cost targets for each main section of the work against the expected expenditure enabling policy decisions to be made before costs were committed in an effort to keep the costs within reasonable limits of the budget estimate.

Fig. 5

MULLARD LIMITED   COST FLUCTUATION REPORT NO............ Date ....

Belmont, Co. Durham

SUMMARY

SECTION OF WORKS	PROJECT BUILDING COST	FLUCTUATION TO DATE		ESTIMATED FINAL COST TO DATE (CONTINGENCIES EXPENDED)
		SAVING	ADDITION	
	£	£	£	£
Civils				
Mechanical				
Electrical				
Direct Purchases				
Claims				
Items subject to separate, additional or re-allocated budget approval				

THE OPERATION OF THE COST CONTROL PLAN

Principal Stages

Whilst cost control is a continuing process throughout the design and construction of a project it can be considered in two main stages:

(i)   The period leading to the formulation of the agreed budget estimate elemental analysis.

(ii)   The period after the formulation of the agreed budget.

The first stage can then be divided into two sub stages:

(i) a.   The period leading to the preliminary budget estimate.

(i) b.   The period leading to the agreed budget estimate elemental analysis.

The Period Leading to the Preliminary Budget Estimate

During the period prior to the establishment of the preliminary budget a number of cost exercises were done to examine alternatives to establish the most economic methods of carrying out certain major parts of the design and also to enable the preliminary budget estimate to be prepared. Some of the most important of these were:

a. Optimising the cut and fill to minimise both the earthworks and also the amount of basement construction which, due to the nature of the ground on certain parts of the site, was expensive.

b. Preliminary designs of the main building roof shape and column disposition to take account of flexibility requirements both of floor areas and roof loadings.

c. Cost appraisal of overhead versus underground conveyors. It is as well to note that not only cost considerations are important but such things as dustproofing, waterproofing, restrictions on flexibility, safety, etc., have to be considered.

d. Various methods of air conditioning the flow coating and gun assembly areas including laminar flow and vertical flow.

e. Total water usage and storage taking account of:

> Cold water
> Hot water
> Chilled water
> Cooling water
> Process water
> De-mineralized water   .
> Sprinklers
> Internal and external fire main
> Trade effluent

These exercises included discussions with the local authority and river authority one of which was to see if it was cheaper for the local authority to treat the effluent. This eventually led to the conclusion that it was most economic for the trade effluent to be treated on site. A point of note here is that this led to the construction of two lagoons, one to hold all the water for dilution of effluent, fire fighting, etc, and the other as a dry reservoir in case of breakdown of the effluent plant.

f. Appraisal of energy supply services to establish economic balance between electricity, gas and oil. This necessitated discussions with supply authorities to define availability and tariffs which then led to a balancing exercise to suit production requirements. Also considered was the security of supply together with the operational complexity of supply change over.

The Period Leading to the Agreed Budget Estimate Elemental Analysis

On this particular project the preliminary budget estimated cost of building and services was considered to be too high by about 10% and it was therefore necessary during the stage leading to the budget estimate elemental analysis to carry out further cost exercises to examine alternatives.

This led to cost revisions to 61 mechanical and electrical services items and 38 building items. The overall reduction at that stage was some £300,000 on the budget estimate.

It is as well to note that not all items examined produced reductions as in a number of instances detailed examination led to an increase.

Whilst it is not possible to enumerate all of these in this study it must be stated that it was not a case of the original budget estimate being over-priced and therefore just a case of re-estimating more exact figures. It was necessary to examine the total philosophy and

73

then look in detail at certain areas where possible economies might be achieved. These were divided into a number of well defined areas as follows:

a. Reductions of scope such as sizes of buildings, areas of roads, vehicle parks, landscaping, internal fittings and all unnecessary decoration.

b. Reduction in specification to ensure that high cost materials were used only where necessary.

c. Reappraisal of loadings on structures and floors.

d. Reappraisal of services distribution to ensure that only those services absolutely necessary for initial production were installed but retaining the main distribution necessary for expansion.

e. Further detailed examination of air conditioning requirements to specialised areas.

Table 1 shows an analysis of the major cost revision achieved without seriously affecting the quality of the construction or flexibility of layout and operation of factory or significantly increasing the maintenance costs.

TABLE 1

	Reduction £	Addition £
Reductions of scope	110 680	
Reduction of Specification	49 380	
Reappraisal of loadings	49 270	
Miscellaneous items		24 450
Reappraisal of Services:		
Mechanical	68 100	
Electrical	47 500	
Reappraisal of air conditions	10 500	
Miscellaneous items		10 500
	335 430	34 950

NET REDUCTION SAY £300,000

It is interesting to note that the percentage reduction of the cost of mechanical and electrical services at this stage works out to 37.5% which is of the same order as the element of total building cost.

The Period After the Formulation of the Agreed Budget

Having reached an agreed budget estimate elemental analysis the cost control procedure must be closely followed if the final costs are to be contained within sensible limits.

There is no magic method of achieving this as it depends on discipline being exercised by all the people on the project, from the Draughtsman through the Designers, Engineers, Contract and Project Management Team to the Quantity Surveyors, Contractors and the Client.

The next major point in the cost development is the issue of tender documents and negotiation of the actual contracts. This is important because it is upon the completeness of the information contained in these documents that the final costs so often depend. It is often difficult when design is being developed to detail this information sufficiently for Contractors to price and understand the work content. This of course is the beginning of the long hard road to claims and increased costs. Suffice to say every effort to ensure definition at the time of preparation of Bills of Quantities and Tender documents is more than amply repaid during the course of the project.

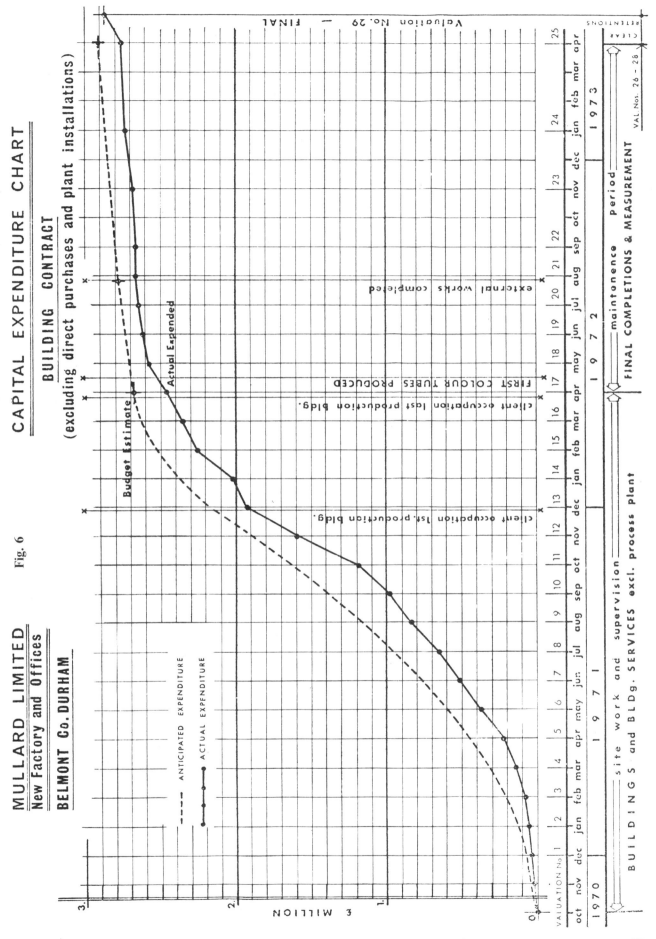

CAPITAL EXPENDITURE CHART

MULLARD LIMITED
New Factory and Offices
BELMONT Co. DURHAM

Fig. 6

BUILDING CONTRACT
(excluding direct purchases and plant installations)

Valuation No. 29 — FINAL

Budget Estimate

Actual Expended

---- ANTICIPATED EXPENDITURE
—•— ACTUAL EXPENDITURE

client occupation 1st production bldg.

client occupation last production bldg.

FIRST COLOUR TUBES PRODUCED

external works completed

£ MILLION

VALUATION No. 1 2 3 4 5 6 7 8 9 10 11 12 13 14 15 16 17 18 19 20 21 22 23 24 25

oct nov dec jan feb mar apr may jun jul aug sep oct nov dec jan feb mar apr may jun jul aug sep oct nov dec jan feb mar apr

1970 1971 1972 1973

BUILDINGS and BLDg. SERVICES excl. process plant

site work and supervision

maintonence period

FINAL COMPLETIONS & MEASUREMENT

VAL Nos. 26 - 28

CLEAR RETENTIONS

75

It is hoped that some measure of the efficacy of these procedures on this project can be shown by reference to Figure 6 and Tables 2 and 3.

It can be seen that for the building contract the expenditure followed the estimated cash flow quite closely. The significance of phased handovers can also be seen by the change in slope of the line.

Table 2 shows an analysis between the agreed budget estimate and the final cost for the various units. It can be seen that the variations are remarkably small considering the complexity of the project. Finally, for those who prefer overall figures, the relative split between Civil Building and Mechanical Services and Electrical Services is shown in Table 3.

TABLE 2

			Cost£/m²	
Indent	Function	Area m²	Agreed Budget	Final Account
A & A1	Amenities, Offices Services Spine and Bridge	6 147	£153.71	£182.02
B	Flowcoat and AC Plant Basement	7 715	£91.60	£111.94
C	Process	4 301	£74.91	£66.31
D	Tubemaking	4 301	£74.91	£66.31
E	Shadow Mask	4 301	£74.91	£66.31
G	Canteen	1 073	£98.81	£72.66
H	Services Block and Switchgear annexe	953	£324.21	£303.33
H1	Process water pump house	111	£73.09	£65.55
	Fire water pump house	28	£49.51	£73.84
	Effluent plant building	229	£69.00	£64.69
	Benzyl Peroxide Store	16	£33.37	£35.31
	Mixed Gas House	4	£23.68	£32.51
J	Inflammable Store	506	£66.31	£76.21
K	Gatehouse	53	£137.35	£141.76

	Agreed Budget Estimate	Actual Cost
	£2 904 865	£2 886 235
Average Cost	£97.63/m²	£97.05/m²
Add Direct Purchases	£251 000	£251 000
Gross Total	£3 155 865	£3 137 235
Cost per square metre	£106.13	£105.48

TABLE 3

Cost Breakdown of Main Building Costs in £/Square Metres

Overall

Final Cost		£2 886 235
Direct Purchases		251 000
	Total Cost	£3 137 235
Total covered usable area		29,738 m²
Therefore cost per square metre		£105.48

Civil Mechanical Electrical Split
   Civil and Building

Final cost	£1 820 955
Therefore overall cost per square metre	£61.14/m²

   Mechanical Services

Final cost (valuation No. 29)	£707,654
Direct Purchases	£189,254
	£896,908

Therefore cost per square metre	£30.14/m²

   Electrical Services

Final cost	£357 626
Direct Purchases	£61 746
	£429 372

Therefore cost per square metre	£13.99/m²

Costs

Included:

1. Trade effluent plant and demineralized water treatment plant
2. Direct purchases
3. Air conditioning
4. External works
5. Main factory services distribution (16 services)
6. Some provision for future extension of services, etc
7. Special partitions, etc, for gun assembly and flowcoat

Excluded:

1. Purchase price of land
2. Professional fees, etc
3. Lockers and cloakroom equipment generally
4. Kitchen equipment
5. Office partitioning and furniture
6. Factory partitioning other than mentioned as included
7. Conveyor and plant installation
8. Traffic lights and road signs
9. Site investigation
10. Cycle and motor scooter stands

DEPARTURES FROM TRADITIONAL PROJECT CONTROL

The main departures from traditional project control at the time this project was started were as follows:

(i) The preparation and use of Bills of Quantities for mechanical and electrical services.

(ii) The use of 'X' Forms to highlight anticipated changes.

(iii) The type and use of cost fluctuation reports.

A number of other procedures were instituted but these were concerned with progressing and whilst they inter-relate with cost control they are not concerned with cost reporting.

The use of Bills of Quantity for mechanical and electrical services enabled a much better understanding of these areas to be achieved. It does, however, require a much greater effort earlier on in the project in these disciplines. It is also made more difficult if the prime concepts are being constantly changed to meet the changing needs of the project. However, compared with the traditional method of using PC sums, there is no doubt that by its requirement for detail early on in the project it imposes a stricter control on all parties.

The use of 'X' Forms, which is now almost standard practice, enables changes to be highlighted even when they are on the drawing board. This gives the Project Management, Quantity Surveyor and Client a much tighter control on costs and enables more time to make decisions before costs are committed. The use of cost fluctuation reports enabled the total cost picture to be seen each month and showed the changes against target areas. Because of the 'X' Forms it reports costs even before they are committed even though they may be approved. This gives an even greater flexibility to the cost management of the project.

It is suggested that these methods, if accepted and worked to by the whole team, get closer to real cost control as against what is all too often just cost monitoring.

THE OVERALL PICTURE FROM THE CLIENT'S VIEWPOINT

General   The Mullard Durham factory was designed to cater for rapid technological improvements in component process requirements and inter-related plant layouts. The factory was required at that time to be built as quickly as possible in order to meet United Kingdom and export sales requirements, the objective of the factory being to produce colour cathode ray tubes of the highest standard of quality at a cost that would remain competitive, both at home and abroad. A total energy concept was seriously considered during the early design stages, but unfortunately had to be abandoned for more conventional systems. The reasons for not proceeding with a total energy system were:

a.   The very important time factor in relation to bringing the factory into production as quickly as possible, and,

b.   The additional capital and design costs involved in a total energy concept.

The factory was designed to accommodate two process lines capable of producing one million colour tubes per annum. This production target required energy and other services to be supplied to the factory twenty four hours per day, six days per week, and the quality of the services supplied had to be of an order of cleanliness and condition that would meet the scrupulously high demand of the process.

The final designs and subsequent installation gave rise to an energy building (Block H—Fig 1) and services complex which generated, conditioned, stored and supplied a total of 17 services to the production units. In addition, considerable attention was given to the disposal of effluent wastes and exhaust gases in order to comply with statutory and environmental requirements. All services except electricity were supplied through the top floor of the central spine and were 'tapped off' as required in the production units. This system has proven itself to be the most economical from a capital aspect and enables production changes to be made quickly and efficiently without detrimental effect on the factory or the energy producing service.

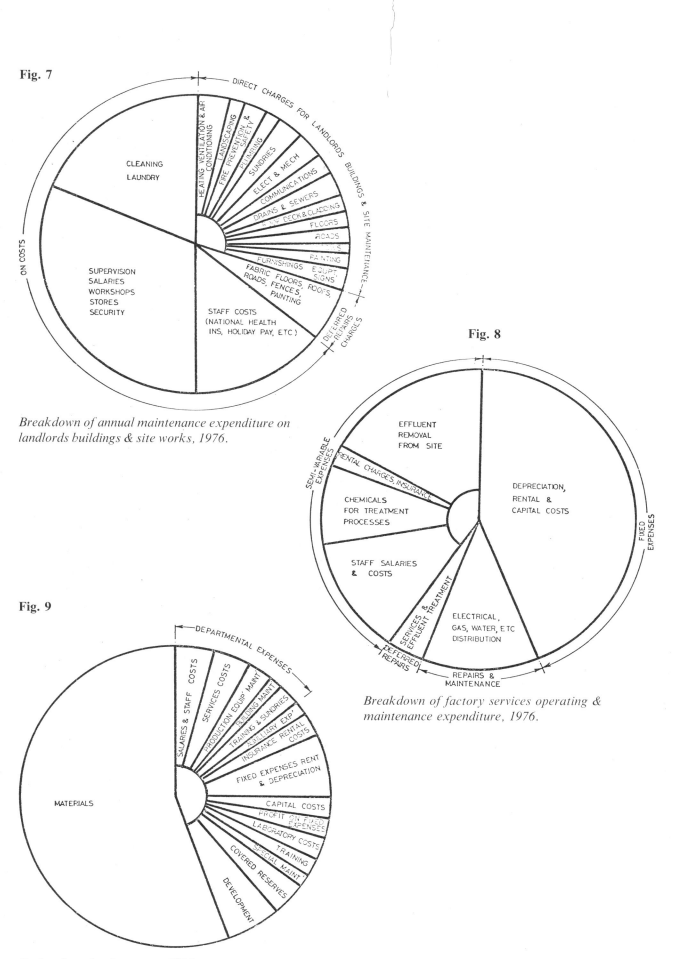

Fig. 7

Breakdown of annual maintenance expenditure on landlords buildings & site works, 1976.

Fig. 8

Breakdown of factory services operating & maintenance expenditure, 1976.

Fig. 9

Ratio of production costs, 1976.

79

Operation and Maintenance Costs

Whilst the cost of purchasing certain services must continue to rise for obvious reasons of inflation, etc., the total cost of operation and maintenance can be continually assessed, monitored and controlled. Operation and maintenance systems continuously assume more complex forms, especially in an industry of this type. The demand for automated manufacturing facilities tends to mushroom costs and imposes an initial requirement in the design stage for a cost conscious approach with considerable weight being attached to the investigation of plant functions insofar as they affect plant maintenance services.

This approach was used in the Mullard Durham factory and the graphs of operating and maintenance expenditure show just how much repair and maintenance costs have been kept to a minimum by good design in the initial stages. At Mullard Durham a distinction is maintained in the accounting system between capital and revenue expenditures, shown in Figures 7 and 8. This method avoids any distortion of the annual budgets for plant and building maintenance in any one year which would be caused by the charge of major replacement works to the maintenance budget. The cost of deferred repairs of this nature are budgeted as maintenance expenses over rolling periods of several years. Typical elements would be roofs, roads, floors, etc.

Figure 7 illustrates the ratio of operation and maintenance costs on all buildings and site works. The heating, ventilation and air conditioning sector appears to be expensive, but this is due to stringent environmental conditions required in certain areas by the process. There is a combination of heating systems used in the factory comprising radiant strip heaters and large heater battery units, of which several are fitted with distribution systems. The radiant strip heating systems are room stat controlled, and the office heating controlled from an ambient temperature control with time switch to give reduced night temperatures. The fresh air requirement for ventilation is tempered by air handling units which have a modulated fresh air control to permit the increase of fresh air to effect cooling when the room temperature exceeds the control setting and the external air temperature is below the control temperature. There is, of course, an equivalent number of extract fans in areas where the air handling units are fitted. These fans give 8 air changes per hour and are fitted with two speed motors which can operate at either speed, both summer and winter, when required by the automatic temperature control system.

The overall heating ratio based on area is equal to 17 therms per square metre, and the volume ratio is equal to 2.20 therms per cubic metre.

Certain areas of the factory are partitioned off into air conditioned clean rooms and carry 10 000 rating under the American Federal Standard 209 (6). This basically means that the maximum number of airborne particles per cubic foot of 0.5 micron and larger, must not exceed 10 000. These controlled environment areas are essential to meet production process requirements and, as such, carry a fair degree of operational and maintenance costs. The air change rate is approximately 35 air changes per hour with a 20% fresh air make up. It is worth noting that the building is generally well insulated and the glazed areas are not excessive. All the main doors are fitted with warm air curtains and/or air locks to exclude the ingress of large quantities of air. The remaining sectors of direct charges are quite normal and are less costly in the Durham factory than other Group factories of a similar nature throughout the world. Included in the total cleaning costs are the 'high level' cleaning costs which are mainly labour costs, and these are due to the access difficulties throughout the plant because of the intensity of overhead conveyor transport systems. As dust is not conducive to the production of good quality cathode ray tubes all high level steelwork is vacuumed twice per year.

It is of interest to note the size of the painting budget. This is entirely due to good design being employed in the building fabric and considerably reduces what is usually one of the largest items in the national building maintenance budget.

The factory is fitted with linear fluorescent lighting and the total lighting load is in the order of 600 kilowatts. The load density of the factory area is in the order of 24 watts per square metre. Twin fluorescent lighting fittings employing reflector tubes have been utilised generally and are located at high levels, i.e. approximately 8 metres above floor level. It is realised that mercury vapour fluorescent or high pressure sodium could be more usefully employed at this mounting height, but the linear fluorescent system was decided upon due to the high intensity of overhead conveyor systems and the knowledge that the conveyor systems would be modified to suit process changes. During the energy crisis period 50% of the fluorescent tubes were removed with no detrimental effect on production.

Figure 8 illustrates the ratio of operating and maintenance expenditure on factory services. It can be seen that costs on repairs and maintenance are quite small and involve mainly labour charges. The majority of the maintenance in this area is of a preventative nature and includes such items as annual boiler inspections, etc.

The factory demands a daily usage of water at a rate of 1250 to 1379 m³. A proportion of this water is de-ionised and passes into the production processes at a maximum rate of 40 cubic metres per hour, at a maximum conductivity of 0.05 micro siemens. All water used is passed back to the effluent plant control systems for treatment before discharge into the local River Wear. The effluent treatment plant is capable of treating 1370 m³ per day of effluent, and treatment is based on a technique of automatic addition of chemical reagents to precipitate insoluble compounds of toxic materials for their removal by settlement and post settlement filtration. In order to operate the treatment plant efficiently it is essential that specified stream separation be effected at the source of the effluent. Therefore a complex system of drainage was built into the factory to cater for water contaminated by toxic metals, lead and fluorides and other general wastes. The cost of operating this treatment plant is quite high in terms of chemical demand and sludge removals. A system is employed of solidifying the water sludges off site by further chemical additives, and resultant solids being available for other uses. This latter process increases costs by 100% but it is felt that it is more than justified in the national interests of controlling pollution. The costs of operating this plant are quite small as the plant has been fully automated and requires just one control operator each working shift.

Figure 9 illustrates the costs of building and services operation and maintenance factors against the total factory production costs. Considerable thoughts and forecast estimation went into the design of the factory and its services and these factors must have played an important role in helping to keep these costs as low as shown.

ACKNOWLEDGEMENTS

The authors would like to express their thanks to Mullard Limited for permission to use various facts in the preparation of the paper and also the various companies and individuals who contributed to the completion of this project.

APPENDIX 1

Project Organization

Client	Mullard Limited
Consulting Engineers and Project Managers	W. S. Atkins & Partners
Architects	A. G. Sheppard-Fidler & Associates
Quantity Surveyors	A. E. Thornton-Firkin & Partners
Main Contractors	Gilbert Ash Northern Ltd.
Steelwork Contractors	Boulton & Paul (Steel Construction) Ltd.
	Robert Frazer & Sons Ltd.
	R. S. H. Engineering Co
Steel Side Cladding and Roofing	R. M. Douglas (Roofing) Ltd.

Metal Windows, Doors	Pillar Patent Glazing Ltd.
Partitions and Ceilings	Brazier & Sons Ltd.
Fire Resistant doors	Mather & Platt Ltd.
Monolithic Grano	Granorite Floorings Ltd.
Suspended Ceilings	Johnson Bros (Contractors) Ltd.
Cross Overs	Durham R.D.C.
Mechanical Installations	Drake & Scull (Engineering) Ltd.
Electrical Installations	James Scott & Co.

APPENDIX 2

Programme Key Dates

Initial Brief	July 1970
Master Plan	September 1970
Project Code Development	October 1970
Site Earthworks Commenced	October 1970
Preliminary Budget Estimate	November 1970
Defined Brief	November 1970
Budget Estimate Elemental Analysis	February 1971
First concrete foundations poured	4 February 1971
Steel erection commenced	1 April 1971
Cladding commenced	5 May 1971
Mechanical and Electrical installations commenced	July/August 1971
Main steel erection complete	August 1971
Main cladding complete	October 1971
Client's occupation of first production building	23 December 1971
Client's occupation of last production building	29 March 1972
First tubes produced by Client	19 April 1972

DISCUSSION

A. J. Colledge (Department of the Environment)

Mr Read in presenting his paper referred to Mechanical and Electrical Engineering (M & E) bills of quantities and said this required a much greater effort in the earlier stages of the project. I would like him to expand on this, did it require a much greater effort in the detailed drawings? A number of points on the M & E bills. Did you have to specify main items of plant and if so how were they determined? Did you go out to pre-tender or just decide yourself and if the latter, were there any objections? Having produced the M & E bills, were there any problems with the M & E subcontractors? Was there difficulty in choosing them and persuading them to accept the bill. Finally following this and presumably later experience with M & E bills, what are your present views?

J. A. Read

Whether to use Mechanical and Electrical bills of quantities really depends on the particular project. What applies to one project is not necessarily applicable on another. When using M & E bills of quantities a very tight discipline must be imposed on the design group in the early stages of the project so that designs are completed to such a stage that bills of quantities can be properly prepared. I think this is a good thing although it does not make it easy for the design teams at the time it is being done.

The preparation of M & E bills of quantities is a problem as, traditionally, most quantity surveyors have no experience in preparing them. The quantity surveyors on this particular project, had tried in the past to prepare M & E bills using quantity surveyors and training them in mechanical and electrical services. I believe they had decided this was not the best solution and on this project they employed mechanical and electrical engineers under a quantity surveyor who supplied the quantity surveying principles. The efficacy of that method was borne out in the quality of the bills of quantities which were produced. The preparation of the bills of quantities is very important. If the staff concerned do not or cannot interpret the drawings and details properly, a bad bill of quantities will result which cannot be understood by the contractor with all kinds of problems resulting. Even a well prepared bill of quantities will not bring isolation from all the usual problems but it will be of considerable assistance in control of the contract both in its cost and through its design.

Regarding the main items of plant, the M & E bills of quantities are similar to building bills of quantities, as they included nominated specialists, prime costs and provisional sums. Major items of equipment or systems which it was essential to be correctly specified to meet conditions within the factory, were first of all designed and then sent out for quotations, separately in order to ensure that correct items were selected. The timing of setting up contracts is important in that the appointment of the M & E contractors came later in the programme than the original appointment of the main building contractor.

In answer to the question on problems of operation. Yes, of course, with the timescale and complexity of this project, there were problems. However, I do feel that these problems were much reduced and then easier to resolve as a result of using M & E bills of quantities. You will be aware that a considerable amount of work is being done on revising standard methods of measurement but until the new methods such as 'method related bills' are brought into practice I believe that for certain types of contract the use of M & E bills of quantities are desirable. Within our organisation we are engaged on a number of projects, some larger than the one described here where mechanical and electrical bills of quantities are being used, and applied sensibly and used properly, they can be a very good tool.

E. S. King (Dunham Bush Ltd.)

Mr Read and Mr Hoskins talked in the paper of total energy having been eliminated from the scheme, but was heat reclaim considered? I also notice that, during the energy crisis, the tubes were removed from some of the lamps. Did this show any appreciable savings? And, finally, figures 7 and 8 show laundry expenses and effluent costs. Would you like to comment further on these. They seem rather large proportions in relation to other costs?

J. A. Read

As stated in the paper we did not give any protracted consideration to a total energy concept due to the very important time factors in bringing the factory into production as quickly as possible, together with the additional capital and design costs involved in a total energy concept. It should be remembered that at the time this factory was conceived there was not the same importance placed on this subject as there is now. It has been considered since, and has been decided that it is unlikely to show significant total cost benefits.

As far as heat reclaim is concerned, this was also not considered at the time for similar reasons as stated above. However I do feel that with the way the services were designed and installed, any changes to implement heat reclaim if it is cost beneficial could be made with minimum disruption and cost to production.

B. J. Hoskins

Introduction of the total energy concept was only considered in a superficial manner on this project. This was entirely due to little internal knowledge of this concept being available at the time and it would have been, I believe, a serious delaying factor in the proposed building programme. We therefore negated this approach firstly from a preliminary capital point of view but mainly from the time factor point of view. With regard to your question regarding the reclamation of heat, we know of definite sources where we can operate heat reclaim systems and we are, at this moment in time, working on this.

With respect to the saving of energy on the removal of fluorescent tubes, I would like to emphasise that I am aware of the fact that, in order to save real energy, it would have been necessary to disconnect the circuits of the lighting system. However, I am not sure whether you realise the difficulties of access in reaching this system because of the various conveyor systems that exist in the factory. Because of this inaccessibility and the cost of disconnecting the electrical circuits, we decided that we would only remove the tubes.

The cleaning costs as shown on the graph seem rather high, but include all aspects of factory and site cleaning plus an internal laundering facility.

The policy within the factory is that in order to maintain the ultra-cleanliness required in certain areas, eg clean rooms, we insist on clean apparel being worn at all times. We therefore, for specific areas, operate an internal laundry scheme. This is also necessary to combat a process problem that would arise from sending overalls for cleaning to an outside laundry. The equipment we use ensures that all laundering takes place in a 'copper free' environment. Copper in the most minute of quantities is detrimental to our process.

With reference to the effluent costs we have to treat approximately 1.5×10^6 1 per day discharge from the factory. The effluent is 'streamed' from the factory into three divisions, general waste, toxic metals and lead and fluoride. The limits imposed on us by the River Authority are quite strict in order not to contaminate the river which is used as a source for an ultimate potable water supply.

We have been removing just under 450 000 1 per week of wet sludge from the site. As a factory, we decided to assist the local environmental problems by paying our removal contractors additional monies to treat the sludge with a 'Chemfix' process. This solidifies the sludge after a period of time and enables it to be used as 'backfill' material, etc. In doing this we have consciously increased our costs by 100%. However, over a period of time we have been carrying out experiments to try and reduce the amount of material taken off site and I am happy to say that over the last few weeks we have been successful. It is early days to reach a final conclusion but the signs are that we can expect a reduction in the order of 75%.

P. C. Venning (Davis Belfield and Everest, London)

Clearly the Mullard factory is a very successful project, which is evidenced by the fact that the client has agreed to come and talk about it. I am particularly impressed by what has been achieved in the limited time scale. I assume that this is the product of close team work and very good communications, and I wonder whether you would like to comment on this, and in particular, I would be interested to know whether the consultants shared an office during the design/construction programme or at least for key parts of it.

B. J. Hoskins

Everybody concerned with the project had to work hard right the way through the project because of the time limits imposed on them. This pressure was placed on contractors, consultants. Mullard Site and Central Engineering Staff. We developed very quickly a feeling of mutual trust and respect despite the difficulties. A point of interest was the unusual factor of our own central architectural division policing the consultant right through the project.

J. A. Read

The point which you raise is an extremely important one and one which is too often glossed over. A parallel situation obviously existed on the Halifax Building Society's Head Office. I must agree with what Mr Hoskins said in his presentation that the development of a first class team spirit of mutual trust and confidence must be achieved very early on in the project.

The setting up of the original team which comprised the client, the consultant and the quantity surveyor, is an absolute essential when you are going to do a job quickly and control the cost, and that team does not stop at client, consultant, and quantity surveyor, a team spirit has to be developed in the contracting organisation and the construction labour force on site.

Another point which is particularly pertinent to the quality aspect of building, is the site supervision. There is absolutely no point whatsoever in specifying high quality and often high cost materials, and then cutting down on site supervision and risking faults arising due to bad construction. This can destroy the whole point of good design or total cost considerations.

I do not think this particular facet of the project can be over emphasised. Unless you have competent site supervisors in sufficient numbers, the client is going to pay when he operates that building in increased maintenance costs and breakdown times. This site supervision is not just confined to the consultants side, it is equally if not more important that all the contractors realise that their site management staffs must also be of sufficient numbers and competent in their respective fields.

A. R. Gilchrist (Scottish Development Agency)

My organisation is the building agent for a Government Agency—building factories to rent, often in advance of a known tenant—and our standards are set with two important considerations in mind. Too high a standard would be criticised as a misuse of public funds, while too low a standard would cease to attract prospective tenants. I would like to ask Mr Read how the base figure was set for this project, since, in my opinion, the cost, as presented in the paper, would appear to be higher than we would provide as our standard. Could you also state the time period used to assess the costs to be set against maintenance?

J. A. Read

The original capital cost figure was derived as follows. Before we were appointed the sales forecasts for colour television sets showed a demand that existing production facilities could not meet. The client therefore had to decide on the viability of increasing his production facilities together with the size of the new development and its location. These deliberations took account of his experience of the factories which he operates both throughout the UK and also the rest of the world. Amongst these cost exercises was included an estimate of the capital costs.

As you will appreciate the final decision to 'go ahead' rests with the main board which in this instance is based in Holland and therefore it was from this decision that we were approached and subsequently appointed to do the consultancy work. At the stage that he was doing these cost exercises ie before he had come to us, he knew far less about how the project was going to develop than we did three months later, after three months of really hard discussion with him. But that was how the base figure was set. That was where the original financial approval came from, saying 'yes'—we want to go ahead to build this job. It is that capital cost that we were presented with.

We then had to try, within that base figure to build the factory, notwithstanding the fact that continuing process development was proceeding during the design and construction periods. It was during the early stages of the establishment of the brief in close liaison with the client, that the preliminary budget estimate was prepared, which followed on from the Master Plan. Once having got the preliminary budget estimate it was 10% over the top of the original capital cost estimate approved by the main board. That is when we had to prune, and that is why the following key stages in this type of development are so important:
   (a)   the establishment of the initial brief;
   (b)   the Master Plan;
   (c)   the preliminary budget estimate (to check whether the financial approvals are being met);
   (d)   the cost exercises (still trying to keep the quality).
As far as the cost/unit area is concerned. I am still amazed when I look at the figures how low this cost is, remembering that this is a purpose designed factory containing a highly sophisticated process, with very stringent air conditioning requirements for the flow coating process together with complex trade effluent problems, and high intensity of mechanical and electrical services to all parts of the production areas.

All of these costs are of course in the cost figures. It is not a speculative development for the type of client who only needs to install a few machine tools to produce his product.

Chapter 6

University of Warwick Arts Centre

N. Thompson

INTRODUCTION

The University of Warwick was one of the new Universities set up in the 1960s, usually on a large open campus site. Some initial buildings were constructed on what is known as the east site prior to the involvement of Yorke Rosenberg Mardall on a larger site approximately half a mile to the west and separated by woods. The various problems and unrest in the Universities at the beginning of the '70s which at Warwick partly took the form of student criticism against the buildings caused the University to decide to employ a number of architects to carry out the next phases of the work. Our practice was briefed to design an Arts Centre and Administration building for the University.

The brief was prepared by the various professional advisers to the University prior to the appointment of the architect. The University envisaged the tacking on of the new auditorium and its workshops to a proposal already in existence for a Studio Theatre, a Conference Hall and other facilities. In our earlier theatre, The Crucible at Sheffield, we had had very detailed discussions with the client before we jointly wrote the brief whereas the brief at Warwick was fairly firmly defined in many areas and in particular the client's priorities. I quote from the brief:

'The University requires that the architects at all times should have regard to the following priorities:

(a) Construction of building within cost limit.

(b) Provision of a building to meet the functional needs of the building as expressed in the brief.

(c) The building to be completed on time.

(d) The aesthetic affect of the building externally and internally having regard to the following:
    (i)   Architectural relationship to existing buildings.
    (ii)   Architectural character of the University and its setting.
    (iii) Internal activities within the building modified as may be necessary, by the system of construction.
    (iv) Such other considerations as may be considered relevant.

(e) Integration of the buildings into a compact environment.

(f) Future maintenance, externally and internally.'

The order of these became changed by a number of circumstances not always within the architects and clients control. The University had intended a 12 month design period followed by a 21 month construction period in order to improve the University amenities as quickly as possible. However before we were able to embark upon our design for the building itself, we convinced the University that the 'tacking on' approach was unsuitable and were asked to carry out an initial study of the whole of the central area in order to establish the best location for the building and its adjacent Administration building together with potential future growth around it.

Fig. 1   *Entrance elevation of the Arts Centre, showing the foyer, roofs, stepping up to the fly tower.*

The contract was let in December '71 to Gleesons for approximately £900,000 including all built-in equipment on a fixed price basis. This worked out at £146·39 per m².

It should not be too difficult for an architect and his team of consultants to be able to design a building to meet the basic brief of the client. This is a relatively straightforward role. What is infinitely more difficult is to develop that brief during the course of the design period so that the building itself fully meets the functional and technical requirements of the brief and provides an economic and viable building easy to run from the clients point of view, but then goes on to provide a number of further pulses: eg options of use within the contained space and cost limit which were beyond the clients basic brief and thus giving him better value for money.

Brief   We are now about to start on site our third major entertainment complex which in fact is in the city of Nottingham. The clients, as at Sheffield, spent considerable time interviewing architects and upon our appointment we were then asked to recommend the rest of the design consultants. The design team and the client then carried out a 2 month feasibility study which led to the writing of the brief which included a detailed assessment of needs and the options open to meeting them and the cost of these options and an evaluation of the site so that we would advise the client properly on whether the site was suitable for the building. The problem at the University of Warwick was that the brief had already been written, the consultants apart from the Architect appointed and the site selected by the client (but subsequently rejected by ourselves) and therefore an important part of the role which an architect can play had already been pre-empted.

Function   An entertainment building normally consists of one or more auditoria, backstage accommodation and service areas, and front of house facilities to provide amenities for the public and allow them to enter the auditorium.

86

The first and most fundamental decision to be made is to define the range of products which are to be placed on the stage and who is to produce them and where they are to be put together. Fashion in theatre auditoria swings backwards and forwards; in the '50s everyone was looking for an extremely flexible space even with large numbers of seats so that uses could range from musicals and evening concerts, medium scale drama and conferences and film and with a wide variety of stage arrangements. The '60s saw a considerable reduction in the range of uses as architects and designers began to see the problem of too much flexibility. Now there is swing back towards multi-use and thus some stretching of the flexibility in order to come closer to an economically viable auditorium from the management point of view. (The National Theatre is an example of a fairly flexible stage arrangement made possible by the use of machinery rather than a considerable number of stage hands.)

Fig. 2

Student route through the foyers, with the bookshop beyond, music centre and conference room to the right and the auditorium on the left.

The amount of backstage accommodation is principally governed by the decision of whether one is going to produce ones own plays on stage and make the scenery for them together with all the props and costumes on workshops on site as in a repertory theatre such as Nottingham Playhouse, Birmingham or Sheffield Crucible or whether one is going to bring in touring plays. One must then examine all the other secondary uses to which these backstage areas can be put. Can workshops be used for non-theatrical purposes; is it cheaper to store scenery in case one might use parts of it again, or is it more economical to not build a scenery and throw it away after a show? Can one let some of the larger spaces backstage for outside uses? Do they make good conference sessions rooms? And what is the cost of providing of this additional flexibility in terms of access and probably improved finishes?

The third principle area is the foyer. Traditionally the foyer was a small cramped space between the box office by the entrance door and the auditorium entrances dominated by staircases and with bars in small corners where they could be fitted in. The new regional theatres being built at the end of the '60s changed this aspect. Theatres became more classless, all seats were entitled to reasonable foyer use and price levels were not segregated out in the foyers. The new theatres were generally run by a trust with a large number of seats on the board going to local authorities and thus became a public place as against the traditional West End commercial manager owner. We set out to take advantage of this situation in the design of the Crucible where the foyers become a public space for 12 hours of the day and provide abroad range of facilities with bars, restaurants, buffets, cold lunches laid out in the upper foyers, New Year's Eve Ball, evening folk concerts and wedding receptions. The success of this and its many implications had a considerable effect upon the design of the Arts Centre for the University of Warwick.

Location

I do not feel that Arts Centres should be regarded as cultural ivory towers. They sould be provided for a broad range of people to go and enjoy themselves and to meet. One should be able to wander in and buy a drink as in any town centre pub. In the process of being there one will meet people who are actually there for a performance or overhear people talking about the performances and in this way one will become encouraged to go to the theatre itself for a production. This relies therefore on people being able to get to a theatre easily and thus its location becomes crucial. Its foyers are the Greek agora, the Middle Eastern bazaar, the Milanese galleria, the English pub. In the design and location of these public spaces the architect is able therefore to get additional use out of the built space and encourage more people to use the central facilities and thus provide the client with better value for constructional cost.

These design aims of fulfilling the broadest possible function and the right location to be able to achieve this range of functional uses within the cost plan makes it essential to have a detailed management study for the whole centre which would include an assessment of the additional revenue set against expenditure for longer open periods, the variable uses of the backstage areas and the effect of the artistic policy which should have been decided upon at this stage. It will also affect the type of finishes that are going to be used in the building and the degree of maintenance which will be necessary because

Fig. 3

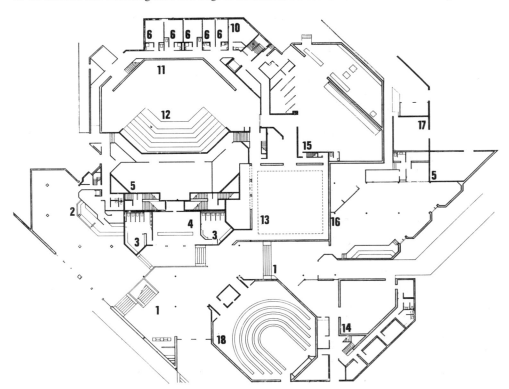

Ground floor plan of Arts Centre (scale 1:750).

obviously if a building is going to be used heavily for 12 hours a day it is very different from an audience coming in half an hour before a show opens in the evening. It is probably reasonable to say that a building which is going to be used for large pop concerts is likely to have a rougher treatment than a stockbroker belt "Arts" theatre.

Growth

Arts Centres are not static. Their artistic directors are creative people who are constantly wanting different ideas explored. If they are successful the demand grows and after a period of capacity use pressure is put for additional building, and improved facilities both technically and socially. We believe that flexibility for the client is best provided by a range of spaces with a limited degree of flexibility within each. This enables the user to select which space is best suited for his purpose, whether it be one of the stages, a large chorus dressing room or area of the foyers with some degree of manual or mechanical modification to the space. Too much modification i.e. flexibility is expensive either in terms of manpower or capital and running cost of equipment and machinery.

DESIGN SOLUTIONS

Master Plan

Our appointment by the University to examine the central area of the University in order to determine the location of the Arts Centre itself led us to question aspects of the YRM master plan. In particular, the planning in which the road appeared to be the master with buildings isolated by space around them and the problem of distance of a quarter a mile between the nearest teaching building to the nearest social and dining buildings with the residencies beyond that. The master plan was therefore basically a three strand approach: residences and their dining areas, a central separate strand of main University cultural buildings and then the strand of teaching buildings. Our diagrammatic plan suggested the considerable overlapping of these areas to provide an informal conglomeration of buildings flanking and surmounting the principle routes and with the public areas of major elements such as the Arts Centre providing the crossing point for a number of these routes.

The Arts Centre is located on the existing main student pedestrian route and serviced by the existing University inner road system. We have also built adjacent to it the main Administration buildings and the University Chaplaincy Centre and these have assisted in being able to define the courtyards and alleyways. A Multi-Purpose Hall for 1200 people was affected by the University cut-backs at the end of the working drawings stage and so has not yet been built and a further area was envisaged by the Vice-Chancellor as a University Art Gallery.

The brief for the scheme from the University set out the function as follows:
"The Arts Centre is intended as the Social, Cultural and Theatrical hub of the University.
To achieve this function it should evolve as a place of interest and character and visually extend and invitation to people.
It should portray warmth and create in the various elements an intimate and comfortable atmosphere.
The Arts Centre will contain the following main elements of accommodation:

1. Performance Theatre
2. Music Centre
3. Theatre/Studio
4. Conference Hall
5. Coffee Shop
6. Bookshop
7. Workshops

and the general planning considerations of the Main Auditorium as follows:
(a) For public performances of plays by touring companies perhaps six times a year, each programme lasting one week.

(b) For public and private performances of plays by the University Dramatic Society.

(c) For musical concerts which could involve operatic performances, choirs and an orchestra of 45.

(d) For lectures as part of the University's academic curricula, to conferences arranged by the University academic departments and public performances.

(e) For occasional public and regular film performances.

Fig. 4

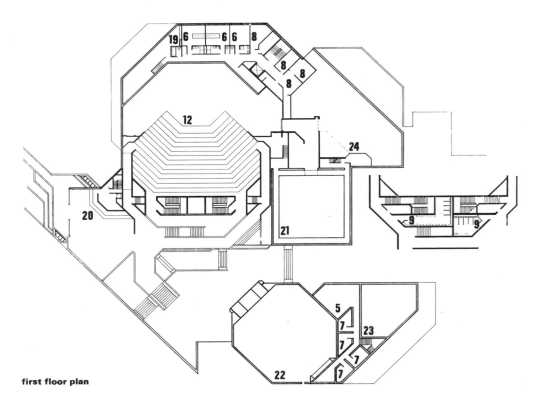

first floor plan

key
1. foyer
2. coffee bar
3. cloaks
4. coats
5. plant
6. dressing room
7. practice room
8. office
9. wcs
10. green room
11. stage
12. auditorium
13. studio theatre
14. recital room
15. main workshop and paintshop
16. bookshop
17. sub-station
18. conference room
19. changing room
20. main bar
21. upper part studio
22. upper part conference hall
23. upper part recital room
24. upper part paint shop
25. upper part auditorium
26. costume area
27. lighting
28. projection
29. dimmer

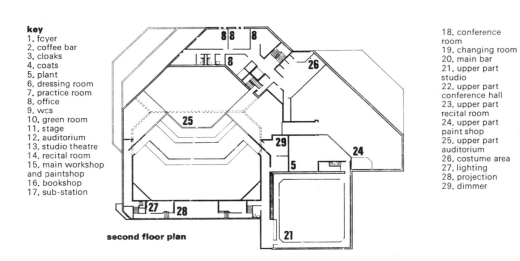

second floor plan

First floor (top) and second floor (bottom) plans.

The group of buildings form the physical and social centre of the University by linking the various residential and teaching buildings which have been laid out by Yorke Rosenberg Mardall. We aimed to provide a focus within the open landscape and allow the buildings to define and shelter both the established and pedestrian routes in dynamic manner in contrast to the previous philosophy of planning isolated buildings in an overall grid layout. Our proposal to build housing alongside and above these routes would further add to the vitality of these principal areas. Although the buildings comprising this group have varying roles they have in common their continuous use by all members of the University and visiting public whether on business, for recreation or just in passing.

Auditoria

The main auditorium seats 547 in its primary form, 577 when the forestage elevator is lowered and three extra rows of seats added. The stage and auditorium are designed to the highest level of regional theatre. Technically, it will meet the needs of professional touring companies, but it is in the first instance a theatre for the University; and thus will be called upon to be used for all varieties of events. They will range from ceremonial academic functions, conference and lecture use and showing of films to full theatical and musical presentations including mounted by the students, using all the technical facilities available.

The auditorium itself embraces both stage and audience, establishing a firm relationship between them by straightforward detailing, and muted dark purples and greys, with the exception of the contrasting graded shades of green to the upholstered seats.

The studio is a much more functional space, providing flexibility at a low cost. The "one room" concept of the acting audience areas permits the breaking down of the division between the two, allowing full scope for experimental theatre. It will also be used by various Faculties (such as the Department of French Studies, History of Literature, etc) for their own particular requirements. More mundanely there is a need for examination space.

The conference hall seats 180 in a high quality interior. Opening directly off the main foyer are the Music Centre, with its Ensemble Room, University Bookshop, bars and box office. The backstage accommodation of workshops, dressing rooms and administration are arranged in 3 levels around the auditoria, with simple functional finishings.

Foyers

The foyers, apart from serving the auditoria, are intended as one of the main all weather meeting points of the University, most students walking through it at least 4 times a day to and from lectures. The route passes all the various activity areas which display information about forthcoming events. The space is designed to express the movements through it and across it; however, it retains a freedom to permit expansion and extension into other associated areas. The foyers are planned as multi-levels for a number of reasons. The entrance level of an auditorium obviously relates to the level of its stage which in turn relates to easy loading from the external areas. This allows one therefore to separate off during the day the upper access levels of this foyer thus saving in staffing costs and allowing us to use the higher quality finishes such as carpets.

The lower levels of the foyers, which are of a size necessary when the main theatre and the future Multi-Purpose Hall are both in use, are however used throughout the day and have much tougher finishes such as paving slabs and Forticrete blockwork to provide the sense of street with people seeing and being seen from the ramparts around cafes, bookshops and other facilities open on to this street. Most of the elements in a complex of this nature require good sound insultation and very few windows. Each space has a different requirement from its neighbour in terms of height, layout and often shape. The building is in fact made out on a meter square grid using 135°, 90° and 45° angles. Our engineers, Ove Arups opted for a steel frame structure which was economic and erected quickly and efficiently and this provided a detailed setting out for the rest of the building.

This structure has then been faced internally and externally in fairface blockwork. This gives it a strength of form which we required for a sculptural building of this nature, provided the walls to the covered street cum foyers and removed fussy and expensive detailing. The large geometric ceiling of the foyer is a light steel frame supported on steel columns, and together with the external wall to the foyer which is clad in coloured glass reinforced plastic (GRP), gives a feeling of a light canopy to this space in contrast to the power of the blockwork forms. The two elements are separated by a narrow perimeter of concealed glazing. We were attempting to use the free natural daylight in a controlled way to wash certain walls and bring emphasis in a theatrical manner to other important areas.

We rejected the down lighter approach to artificial lighting and instead went for the very powerful arrangement of fluorescent tubes, switching of which was organised to allow us to vary the levels of illumination depending upon the use of the building. But probably the most dramatic use of the foyer is Clare Ferraby's zany use of colour variously described as *stepping from the English countryside into Las Vegas*, to *an explosion in an Italian ice-cream factory*. The crucial thing is that it is talking point that stimulates people, arrests their eye as they walk through and is achieved at no extra cost whatsoever to the client over using white emulsion paint. Equally in ten years time when everyone is tired of it, it can be repainted in whatever way someone likes, and it will change the character of the foyer.

Backstage areas are built to a very basic standard of exposed blockwork, pvc floors and exposed trunking and services. We believe that theatre clients are more interested in having as much space as possible for their money rather than elaborate finishes and exposed services certainly allow easy modification including the use of a space for a different purpose than that originally planned.

The main auditorium is fully air-conditioned with imput at high level and extract in the void under the precast seat risers. This is partly in order to be able to attract good summer vacation conferences.

Elsewhere, services are simple; with radiators backstage, and a few fan convectors in the walls of the foyers. The University high pressure main provides the heat source, to a central calorifier and then to a number of local air handling plants, thus reducing ducts and the need for false ceilings. Too often spaces of this nature have too sophisticated services. The sub-contract for bringing in the heating mains, ventilation, air conditioning, with its associated electrical and builders work, including main contractor's preliminaries, was 10% of the total cost.

The external spaces have now been heavily planted and we are persuading the client to grow creepers over many of the walls. It is probably one of the few materials one can think of which increases in value over the years, changes colour over the seasons and welcomes sun and rain. Furthermore the people who maintain these particular areas seem to enjoy it and look well on it.

CONCLUSION

The clients first priority was the construction of the building within the cost limit. This meant that we were forced rigidly to adhere to the contract sum of approximately £900,000 which worked out at £146·39/m² including all the specialist services required in a building of this nature. It is possible to keep this level of costing provided one uses extremely simple materials which are selected to perform a wide range of functions without the introduction of numerous secondary materials. However, they do need to be offset by materials of a more glossy and luxurious nature where one comes into contact with them, and in the Crucible this was achieved by an additional grant towards the end of the job to upgrade such items as handrails, doors and other elements which one was physically handling. This was not the case at the University of Warwick where once we

Fig. 5

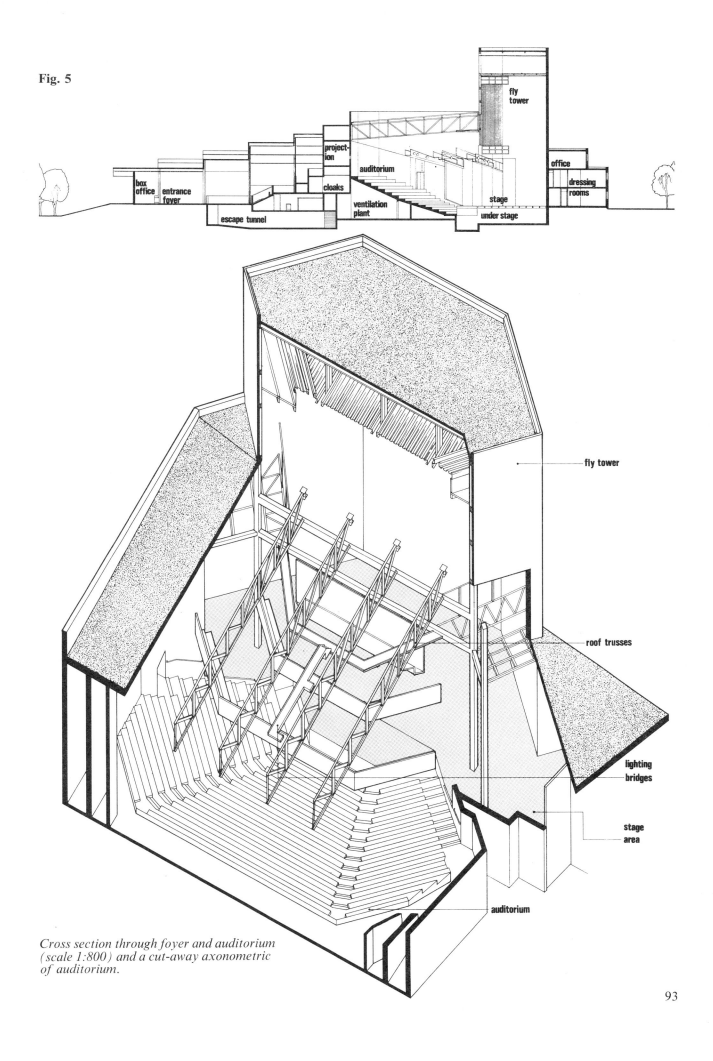

box office

entrance foyer

escape tunnel

project-ion

cloaks

ventilation plant

auditorium

fly tower

office

dressing rooms

stage

under stage

fly tower

roof trusses

lighting bridges

stage area

auditorium

Cross section through foyer and auditorium (scale 1:800) and a cut-away axonometric of auditorium.

93

had reached our relatively small contingenices level we had to make a saving in one part of the building to offset any extra that was required elsewhere. This seems a false economy as already the client is replacing some of these omissions or reductions with more expensive materials of the type which we originally recommended. The maintenance of the simple finishes which we have used and the cleaning of them appears to be low in cost. It is also interesting that at a University where it is normal for students to deface the buildings and particularly to cover them in posters and stub out cigarettes on the carpets there does not appear to have been any wilful damage in this particular building.

The illustrations are reproduced by kind permission of the Architectural Review.

Chapter 7

An Objection to Tero-technology

B. Drake

Buildings last much longer than most machines and techniques which are helpful in engineering often require adaptation for use in building. The present worth of reducible maintenance costs is small compared with capital costs so little money is available to improve maintenance characteristics. The future is unknowable and is largely not worth knowing because of the effect of the discounting function. It follows that excessive devotion to the theology of terotechnology is misplaced in a building context. The essence of terotechnology is a holistic approach to product design; we are urged to consider the entire life cycle of a product and not to concentrate our attention narrowly on initial design or capital cost alone.

Our predecessors in building have a pretty bad record in this respect. Certainly the man who costed St. Bartholomew's Hospital 800 years ago and whose building is still in use was reprehensible in failing to make provision for cardiac surgery or renal dialysis units or even for the regular use of an EMI scanner. But perhaps his record can stand comparison with others; I believe the designer of the Parthenon made only very limited provision for the parking of tourist coaches and the Egyptians have had to build special pavilions for the sale of postcards at the Pyramids. All of course most regrettable lapses but reflecting perhaps the truism that buildings last rather longer than most machines and that in consequence techniques which no doubt make sense in engineering require adaptation before they are likely to be useful in building.

Moreover the importance of maintenance costs when compared with capital costs can easily be overstated where buildings are concerned. Bathurst and Butler[1] have shown that the annual equivalent cost of repairs and replacements represents about £86 per 100 m^2 against an annual equivalent of capital costs of some £602 per 100 m^2 (discounting at 9%). However much of this £86 is attributable to renewing plumbing, heating and electrical services every 30 years and to regular re-lamping of light fittings. It seems unlikely that higher initial capital investment could do much to reduce or eliminate such renewals. If therefore we ignore these items we find that repair and replacement costs have an annual equivalent of about £42 per 100 m^2 or about 7% of the annual equivalent of the capital cost.

So we have to ask ourselves whether an additional capital expenditure of 7% could totally eliminate all repair and replacements, all washing down and all redecoration throughout the life of a building. In other words no roof would ever leak, no lock ever fail, no floor finish ever wear, no door ever be damaged and so on.

To state the problem is to demonstrate the absurdity of the concept but it is on just such absurd premises that Government is invited, even urged, to either abandon capital cost limits entirely or to substitute a system of cost-in-use cost limits. The former proposition is clearly inconceivable even though there may of course be a few private clients wishing to indulge whims, their own or their architects! The latter proposition, cost-in-use cost limits, is altogether more respectable but as we have seen maintenance costs are only a very small proportion of capital costs and we would not be justified in heavyweight cost-in-use procedures to incorporate maintenance cost alone. And they would be heavyweight; we would need life profiles of all materials and components and these would have to be officially established and promulgated. These profiles could not be factual because, as an unpublished paper by Azzaro of DOE (PSA) has demonstrated, over 70% of a typical PSA building, say an office block, would be designed or built differently now from the way it would have been designed or built only 20 years ago. It follows that feedback systems on component and materials performance would be of little value in cost in use calculations; of course they have other uses.

Moreover we could not accept a simple approach whereby a design team made a case on cost in use grounds for, say, a roof finish requiring additional capital expenditure, because it is quite possible that a cheaper solution elsewhere in the building would have allowed them to afford the roof finish without further capital. So we would develop a substantial bureaucracy poring over consultants' cost-in-use calculations and no doubt squabbling with them; scarcely an enticing prospect. As an aside it is worth bearing in mind that, at present discounting rates, cost-in-use calculations often indicate the use of lower quality components.

Now it may be objected that all these arguments are excessively dismissive of a serious issue and that no examination has been made of the question of heating costs, cleaning expenses or other operating charges which can amount to substantial sums. But whether these issues are important or not turns on a forecast of future heating or cleaning costs etc. and forecasting is a tricky business as Newbold[2] reminds us:

"Many of the quantitative techniques in current use in forecasting involve highly sophisticated model building exercises. The models employed are based on up-to-date theory, large quantities of data and often a highly complex statistical analysis. Nevertheless, when one considers the numerous ways in which outside events can upset the calculations of any such model, the quantitative analyses appear almost trivial by comparison."

An observation, one may feel, which applies with particular force to the time scale of buildings during which not one but a number of complete social, economic and technological revolutions may occur. In the face of such colossal changes, an obsession with, say the prediction of the cost of cleaning throughout the life of a building when the future relative wages of cleaners compared to other members of society is quite unknowable, seems a trifle lacking in balance. Again, there can be few calculations of fuel costs or optimum insulation thicknesses made even as recently as 3 years ago which are now in any way valid.

A further discouragement to peering into a misty and uncertain future is the effect of the discounting process. The sums are of course well known but it may be worth recalling that the present worth of £1 payable in only 20 years time is–

discounted at 5% = 37 pence
discounted at 10% (test rate for nationalised
                            industries) = 15 pence
discounted at 25% (1976's rate
                            of inflation) = 1 penny

The effect of discounting can scarcely be ignored; especially since it seems to agree with common sense in that it has always seemed a reasonable proposition that each generation should solve its own problems. Though one would perhaps admit that we should not go out of our way to make life difficult for our successors, the probability is that they will be richer than us, in which case what are we worrying about; but if by chance they should be poorer, our buildings will be an embarrassment to them and will need to be significantly altered or adapted if not demolished. So again an undue concern with the minutiae of terotechnology seems misplaced. Well this is all very unconstructive and it is no doubt time to summarise one's position.

If the exponents of terotechnology are trying to suggest that the millenium awaits some knight on a white charger who will slay the dreaded dragon of bureaucracy and release the fair maiden Design from the bonds of capital cost limits then one must beg to differ. In fact the relative costs of maintenance and capital are such that very little money is available to improve initial design and certainly nowhere near enough to eliminate maintenance charges. It is likely that careful attention to detailing at no extra capital cost would produce better results in maintenance savings than would increased capital expenditure.

Forecasting accurately over the life of a building is impossible, fortunately the effect of the discounting function is to make any event occurring more than 30 years after completion of no financial significance to us—and *we* must make the sacrifices and the decisions. So we need only make our buildings reasonably adaptable (and we would not be justified in spending much extra to do so) and we need only solve those problems which face us today and those which we are confident we can foresee whilst trying to leave options open to solve future problems (roof spaces in which we can lay insulation, for example) provided we don't spend unnecessarily in doing so.

None of this is to say that it is not sensible to consider ease of access or cleaning or to reduce or eliminate maintenance operations or provide for adaptability, where this can be done at small extra cost. But an undue attention to terotechnology theory, an over elaboration of its methodology or an unreal attempt to foretell events 50 years hence seems to fly in the face of common sense and to invite ridicule from our successors. In short, it is not more money we need it is more wits.

REFERENCES

1   Bathurst, P E and Butler, D A. *"Building Cost Control Techiques and Economies."* Heinemann. (Table 18.1).
2   Newbold. *"Forecasting Methods".* Civil Service College Occasional Paper No. 19. HMSO.

DISCUSSION

D. M. Andrew (National Westminster Bank Ltd., and British Council of Maintenance Associations)

I must thank Mr Drake for a witty and interesting paper and for such pleasantly put objections. I am sure we would not think 30 year projections to be possible but what is going to happen in 3 or 4 years' time? We are now thinking in terms of inflation accounting and the Sandilands Report.

Terotechnology is management applied to physical assets, in pursuit of economic life-cycle costs; we have to look at these annually. I am worried by Brian Drake quoting the pyramids with such colossal U values. But people do allow space in modern buildings, for example, the Bank of England was refitted 2 or 3 years ago with new lifts which were fixed to RSJ's so that in 30 years' time, if a new and better installation needs to go in, it can go in quite easily.

The subject of Terotechnology is supported by the Industrial Technology Division of the Department of Industry who have helped tribology, materials handling, and research on corrosion. However, they will only give it a boost for 4 or 5 years, spread the interest, and then leave the subject to stand or fall on its merits. We feel that Terotechnology really means a greater involvement of the client, the property manager, the engineer, the financial controller, and the fire security and telecommunication specialists with the architect, the technical consultants, the quantity surveyor and possibly services consultants, if the job is big enough.

We would underline that the reason why Terotechnology is important is because of the need for speed in commissioning. In addition, the increased population densities in city centre offices and the higher environmental expectations of these populations are increasing energy problems. There are increasing security problems, which affect our telephone, fire and ventilation arrangements and which, for instance, bring in precautions against vandalism we did not have to consider even 10 years ago. There are interface problems—we need the systems and electrical circuits flexibility indicated by several speakers—the use of options engineering, the provision of dual fuels, etc.

I saw recently the annual running costs for the Greater London Council's new 7-storey Island Building near Westminster Bridge. They are half the cleaning costs in the main County Hall building which is entirely due to maintenance design improvement. We are only interested in the art of the possible. Terotechnology is not a new science, it is the team-work concept, which will only work if all the members of the team are equally consulted and equally involved.

R. Cullen (Architects Design Group)

Jack Coia said at a lecture I went to a long time ago "Find out about the past, get on with the present and forget the future. Stop worrying about it". I found myself agreeing with everything Mr Drake said and yet could not agree with him. When you think of our cities as they are now and you look at the buildings that make up those cities and make one city different from another city, you will find it is the buildings at least over 50 years old, if not 80 or 100 years, which make the cities we love and enjoy. I would submit therefore that we need to spend as much money as we possibly can on the buildings we are building now. And although the RIBA may have got it all wrong and, according to Henry Swain, are saying it for all the wrong reasons, I think they are saying it for the right reasons, although they may be intuitive.

B. Drake

It would be very insensitive not to take your point but I would like to make a further analysis. You are quite right I think about cities, although some of the attractive areas of cities were tolerably working class areas at one time and are now being 'gentryfied'. In the countryside though there are many buildings which were originally extremely economical. A later paper demonstrates just how well adapted those buildings were. I am not convinced it is necessary to spend a great deal of money to produce a building which will remain admired over a long period of time. The peculiar dynamics of city development, and the absence until recent years, of planning laws may have had more to do with the wiping out of the smaller and less monumental buildings in our cities than did initial capital expenditure. However I put this forward only as a possible hypothesis for testing. The general view propounded by ministers, newspapers and opposition alike is that greater investment is needed in factories and manufacturing industry. That view may be misguided and it is an entirely honourable occupation for people to object to it and offer alternatives. Whether a richer society should be prepared to spend more on its buildings is a moot point.

Professor Douglass Wise (Institute of Advanced Architectural Studies, University of York)

We are involved in York at the moment with problems of conservation and re-use of existing building stock. In fact the Institute occupies a building started in 1281 and which has had nearly 700 years of alterations, adaptation and addition, and still meets a modern need exceedingly well.

We must not delude ourselves however that all the buildings which we regard as our heritage were a cheap commodity at the time of building. Some artisans buildings happen to have survived, such as the Shambles in York, but a great deal of our heritage which we treasure and take pride in was probably very expensive to build. I would guess that York Minster and St Pauls were much more costly relatively than the National Theatre or Sydney Opera House. We must not draw too many general conclusions from our building heritage.

D. K. Barbour (The Riches and Blythin Partnership)

Conservation of land has been discussed. We talked about cities and city development, about building and cost of buildings. It is my view that we ought to be considering now a much more fundamental change in building forms and types; something which might not really end up as buildings at all. There is no technical reason why we should not take a space, put a huge dome over it, air condition it and control it. Everything city dwellers need to live and work could be put inside. By generating air conditioning on a vast scale and using detailing materials unconcerned with weather we could achieve enormous cost benefits, flexibility and land conservation.

B. Drake

You are much more confident than I am of a knowledge of the physiological responses of human beings in such a situation not to mention the psychological ones and working in an air conditioned office all day can be a very monotonous environment. I would like to know a lot more about that fairly elementary side of things. There are also problems of fire and noise which are not quite as simply handled as you seem to suggest. There is another matter which concerns me rather more though. We have had a poor experience of what happens to exciting and interesting architectural concepts over a period of years. Ville radieuse is one thing and what actually happens in half the boroughs of the North East another; and all those little country towns in Scotland—the last thing they needed was the eleven storey tower block they now have.

E. J. Boyle (Haden Young Ltd.)

I would just like to investigate the question of looking into the future. If a major feature of Terotechnology is to look at the future, I doubt its usefulness in long term planning because forecasting events many years hence is almost impossible, if it attempts too much detail. Maybe the plastic dome is a solution but equally we could all be sitting at home with small computer terminals doing our work in thirty years time. It would be better if we concentrated on doing our job as we know it, much better, looking at maintenance, and energy conservation of course, but confining our long term view of the future to the problems of the buildings we currently require. Does Terotechnology help in this respect?

J. M. Cooling (Institution of Heating and Ventilating Engineers)

Mr Drake commented on his experience working in an air conditioned office all day. I wonder whether in fact his office is air conditioned or whether it is one which was designed as an air conditioned office, which has not been maintained and is therefore not an air conditioned office any more.

B. Drake

The standard of maintenance is good. An engineer might catch me out on definitions of air conditioning, but it is a reasonably advanced system. I think everyone finds that sometimes when one goes out there is a sense of exhilaration although the reverse happens too, of course, in muggy weather. If you are twelve hours on a Jumbo Jet do you not try to get to the door as fast as you can to get a breath of *fresh* air.

H. S. Staveley (Martin Staveley & Partners)

I have spent what amounts to very nearly a lifetime suffering from having to look after the maintenance of buildings that had been put up on the *ad hoc* design and construction principle which Mr Drake is advocating. I do not believe him because in effect he has to a great extent been advocating the principles of Terotechnology. There is only one little bit of the Terotechnology process he does not like and that is the discounted cash flow application. As far as the rest is concerned he is with us 100%. This has been proved by the fact that his Department is busy constructing and have constructed several hospitals which they refer to as the 'best buy' hospitals, these being intended to last for a minimum of 60 years of life.

R. J. Wilde (Westminster City Council)

A previous speaker mentioned that we cannot forecast for ten years time. I would hazard a guess and say that unless we do something now tenants will still be sitting in flats heated to 60 °F and complaining about the bathroom fan noise simply because we have not practised a tiny bit of Terotechnology such as telling the fan manufacturer to spend a little more money on a more robust design and make it last for 20 years instead of 12 months. A good example is the stainless steel car exhaust system which for a few more pounds will last the car's lifetime against 2 years for the standard fitment. This is all Terotechnology and means the saving of money in the long run. The Committee on Terotechnology formed a few years ago estimated that £1000 million per year could be saved in British industry by practising Terotechnology. I think we are missing the point, we do not wish to spend more money on maintenance, we wish to spend the money we have to better effect.

Mr Drake said the engineering services are becoming increasingly sophisticated and for that reason require more attention. Perhaps we should look hard at this situation and design systems which are simpler and can be understood by average operation and maintenance staff.

Referring to my earlier remarks on under-heated flats. We are still designing to Parker Morris standards which are now out of date, should these not be revised? If we adhere to the Department of the Environment yardstick it would seem that we are only asking for trouble from tenants in the future.

B. Drake

This is a very valuable and necessary corrective to some overdone stuff, but does it really cost more money to produce a quieter fan? Perhaps it does. I would argue first for buildings which are adaptable, for example, do not chose an optimum instulation thickness and say that is that, you see if you can find ways of making it flexible, such as the roof space of the ordinary domestic house where you can put almost any amount of insulation. Could we not make life a little easier by employing Alex Gordon's philosophy of long-life, loose fit, low energy.

P. E. Bathurst (North West Thames Regional Health Authority)

I just wondered if I could join with Mr Drake to make a final comment on this point. We have had something like 25 years working in the government service setting cost limits, designing schemes within cost limits and devising procedures to operate cost limits. We did this because it was what our political masters considered to be necessary and what we are saying is quite simple. If we went back to them and said we want more money for buildings we just would not get it. This does not mean that we have not learned about the cost of maintaining buildings or the importance of building to proper standards. What we are saying is that it is much more a matter of making the right decisions. Mr Drake says we have got to use our wits better.

The target at which we were both shooting is the automatic assumption that the expenditure of more money is going to produce these results. We could instance many examples where the extra expenditure necessary to save maintenance in the future has simply been wasted on uneconomic planning or extravagant detailing. These are the things which have absorbed the money which could have been devoted to reducing the incidence of maintenance.

When you consider the Department of the Environment housing yardstick there has been, for many year, the possibility for a local authority to spend more than the yardstick and where it could show that maintenance would thereby be reduced, it is not the yardsticks which prevented people spending money on maintenance. I believe that both Mr Drake and myself have been making this same point that you have to use all the systems of analysis and methodologies to consider and decide what are your priorities and how you are going to achieve them. The automatic assumption that more money is necessary is not the right response; and I would like to present two examples.

Quite recently, when the new regulations were introduced for thermal insulation, there was an immediate reaction from local authorities and members of Parliament, that this would clearly mean more money must be spent on housing. When the standards of specification actually used in new housing were analysed it was discovered that most local authorities were meeting these standards and so for the majority no extra money was required. This was far from being the general view as expressed in the architectural press, or by the industry where the universal reaction was that it was both impossible and expensive.

The other example is of a physics building where there were complaints about the windows—noise nuisance from one room to the next via the windows, glare from the high exterior lighting level, solar gain etc. finally there were complaints that the University Grants Committee would not allow capitalisation of running costs to have double glazing. Yet every other complaint made it quite clear that the trouble was that the building had about three times too much glass in it. The occupants wanted the University Grants Commission to have a system which would allow them to capitalise on running costs. They might just as well have asked for a system that prevented architects from putting in more glass than was necessary. The point of this example is, therefore, what kind of control do you really want on your buildings? I fear that if you insist on having total cost limits, control of specifications will be the inevitable result.

H. G. Mitchell (H. G. Mitchell & Partners)

If Part F of the Building Regulations is extended into buildings other than domestic ones, architects are going to find that they will have to reduce the glass area.

Professor Douglass Wise

I feel slightly responsible, as a representative of the architectural profession, to correct the record. Mr Mitchell knows very well that the movement to reduce glass sizes, or at least understand the problems involved in using glass, started in Newcastle in the School of Architecture and was the result of research work done on the relationship between insulation and fenestration. It is also very easy to have perfect vision in hindsight and to be highly critical of past endeavours. As an industry we fell into all sorts of traps which we can maybe see now clearly, but which at the time were infinitely more obscured.

R. Cullen

I am going to be deliberately provocative. I think you could be not acting professionally and responsibly in accepting the politicians' 'What's left over'. Now the building industry, everybody tells me, is suffering and will go on suffering until it is no longer the recipient of what is left over. Planned programmes and planned expenditure are needed and not subjection to the ups and downs of our mad economy. Surely it is Mr Drake's job to tell the Government just that?

B. Drake

Clearly we must tell the Government of the consequences for the industry of the fluctuations you describe. I have only been in a position to do that—and doing it—since 1969. We were ignored until 1974. I think we have been listened to since. There is now a Standing Committee which looks at the load on the industry regularly and it has a reasonably well developed forecasting system. However, I would be unlikely to be able to stop a future Chancellor of the Exchequer, with an election to win, injecting money into the industry too fast. It does not seem to me I am being asked to act unprofessionally if ministers choose to ignore my advice.

On the other issue of how much should be spent on buildings I must disagree with Mr Cullen quite totally. I can advise ministers as to the consequences of various levels of expenditure on buildings and I can try to make that advice as broad as I can. I can try to embrace aesthetic values and social consequences, financial consequences and economic consequences for industries, and consequences for professions etc. I can do all that but I cannot and neither can architects take on themselves the responsibility for making the decisions about how much of the money available ought to go on buildings, such a stance would approach arrogance and where it has been detected by the public it is resented.

I doubt if the architectural profession has known what is best for people. High rise flats were purely an architectural concept, land values hardly entered into it, it was a following through of an ill thought out, inadequately researched, inadequately argued architectural concepts of the 30's which has been socially disastrous. Any profession must do its best to construct a dialogue with the public and preferably an informed public, but if it starts from the premise that it knows best, it may be in for a shock. I have some concern and this is shared by public concern, regarding the ability of the architectural profession to handle the scale of modern life. I think it has defeated them and that of the many comments made about bureaucracy, whilst some are justified, others may be borne of a frustration at the inability of the profession to handle the scale of contemporary demands. I do not pretend it is an easy problem, I think it is an immensely difficult problem but I am equally certain that the architectural profession has not learned how to handle it. Until they do they will not regain public sympathy and still less will they get it if they try to impose their views on the public.

R. Cullen

I hope I am not arrogant but humble enough to learn from others. Of course Le Corbusier had a utopian dream and utopian dreams like domes are very dangerous things, but what created the high rise boom in this country? It was not architects. It was a combination of circumstances. It was speculators. It was surveyors who said there is a bit of land which is so much money, and if we are going to make a profit you have got to have a block 16 storeys high. This attitude and calculation produced buildings which we, the architectural profession and the public have now realized are unsatisfactory. Another major contributor to the malaise was a Ministry booklet 'Mixed Housing Development' published in about 1954 which promoted high and low rise mixed development.

B. Drake

I partially absolve architects from offices but I do not from housing. For offices neither the valuation surveyors, advising the property developers nor their architects, got it right. Nor of course would the quantity surveyors have done either.

Professor Douglass Wise

I am half an academic and half a practitioner and the academic half has a certain respect for the truth. There is a growing volume of knowledge, based upon research, both scientific and sociological, which is beginning to indicate that our reaction to the high rise flats may not be objective, and that they can have many things in their favour, particularly when compared to the serious problems of some low rise, high density complex developments.

Brian Drake, however, is I think quite right in one thing. I do not think we as an industry can handle some of the big problems. But it can be put another way round. I think we have been presented with problems incapable of solution. A great fault of the architectural professions in recent years is that it has assumed that every problem had a solution, or at least a partial solution. This was the result of training in project based education and a pride in being able to use our skills as problem solvers. We might have been wiser if we had said to society—restate the problem, think it out more clearly, or break it down for as it stands it is impossible of solution.

T. Smith (Steensen Varming Mulcahy & Partners)

With some regret may I endeavour to bring the discussion back from the socio-political level to the mundane? I suppose it is inevitable that money must be the factor of measurement in all discussions relating to Terotechnology. We have spoken of the possibility of increasing capital expenditure in order to reduce revenue costs. May I suggest that there are two other facets which are worthy of consideration when we consider the economics of Terotechnology. One is the question of time and the other of energy.

Time, I suggest is more valuable than money although they are essentially linked. In the phrenetic activity described as 'the design process' we do not today have sufficient time. If we add to this present activity an honest desire to investigate all aspects of Terotechnology before making our design selection, I suggest that the time available requires to be dramatically increased.

In respect of energy there are two forms, firstly human and secondly natural. It is the latter to which I wish to refer; that is the use of natural resources for the generation of energy to manufacture building components, to transport them to site and to erect them as part of the building. It is necessary that we appreciate that the rapidly depleting energy sources of the world mean we must give much more thought to all these aspects of energy usage than we have traditionally done. Again, such consideration requires time and time is money. The cycle has, therefore, fully turned and we are back at the point at which I commenced.

Professor P. Burberry (University of Manchester)

I would like to refer back to the subject of high rise buildings. During the course of the conference, I have heard with increasing alarm a reiterated statement that these buildings were forced upon designers by economic necessity and by planning legislation. Both of these statements are clearly incorrect. High buildings are expensive in their nature. They could have been forced upon us economically by high land costs but the planning legislation defines a ratio between the site area and the amount of floor space allowed which is sufficiently low to mean that it is never necessary to use a high rise building because of site limitations.

It may be that architects prefer the appearance of the high rise, rather than the lower development but this is something quite different from being forced to build high. In thinking about this point, it is important to look back to architectural statements made in Europe just before the War when high buildings were very actively advocated ostensibly because they gave fresh air, freedom from noise and adequate lighting. A great deal was written on this subject. Although it was not a statutory provision, the approaches to daylighting in town planning were modified in order to allow high buildings. Unfortunately, the reality which was discovered was that as buildings increased in height, we did not get the fresh air because it was discovered that stack effects and wind effects were such that buildings had to be sealed and air conditioned. We did not get freedom from noise because although the effect of the adjoining street was reduced, noise from distant streets made a contribution. Finally, beyond the protection of other buildings and trees, exposure to the uninterrupted sky gave glare and solar overheating.

If all these things had been unpredictable, the situation might be readily excusable but, this was far from being the case. These effects could have been predicted using knowledge available at the time. If this was not regarded as satisfactory the whole circumstance could have been observed at work in the United States. Nobody in Europe appeared to think that this was necessary and now in Europe we have a vast amount of high buildings which, whatever their merits might be, had their origins in serious misconceptions.

C. K. Robinson (University of Manchester)

I am one of the generation that came down in busloads in the '50's to see what the London architects were doing in the way of local authority housing. The argument which stuck in our minds as being the most salient reason to build high and to build with that separation was the ground that one gained in between the buildings was of a sufficient scale to support and allow these large concentrations of people to live in a sane way. I can remember being impressed as a young man by the seemingly reasonable argument. At the top of one of the Roehampton slab blocks the job architect explained to me the need for pressed metal baths and pressed metal door frames in an effort to stop vandalism. I am a Mancunian from a campus close to one of the biggest deck access developments in western Europe. It is an utter failure.

Many prophesied that it would be a failure because one of its 'principal attractions' was, for no good reason, to get people to go up three storeys or more and then walk over many, many acres, blown about by the wind, over streets and roads whilst up in the air. Now the city fathers are to ask central government for several million pounds to take out the deck access bridges. I wonder what Mr Drake will say in ten years time when he sees the results of the reaction to high rise buildings. In Manchester the reaction is disastrous; high density, low rise developments which are impossible to police and into which it is impossible for fire brigades to go, since when there are fires the firemen cannot get water through before the youngsters put knives through the hoses. It is very easy to malign high rise and it is easy also to malign the low rise developments but I do wonder what we can actually afford to build.

B. Drake

May I take up the two points; time is more important than money, and what was really a plea for energy budgets. It would take far too long tonight to explain why I am opposed to energy budgets. I think that the financial mechanism is the right one to use. If you go in for energy budgets you are in danger of distorting the use of resources. As regards time being more important than money, most people employ money as a proxy for resources and usually it is a fairly good proxy. I would have thought that provided your equation is comprehensive and includes design time and thinking time etc money could continue to be used as a proxy for resources.

Chapter 8

Optimisation Techniques for use in the Design of Buildings

M. Carver

INTRODUCTION TO OPTIMISATION

Criteria

When a decision is taken to optimise a building design the question arises 'What is to be optimised?' The criteria by which solution is judged to be the optimum must be defined. It is possible to optimise human comfort or satisfaction, but there are not universal scales whereby one individual's degree of satisfaction may be measured against another's.

Where there are several criteria by which the optimum is judged the probability of any one solution being the best of each criterion is remote. In practice therefore, one criterion is usually adopted and this is based on the use of money. The objective then becomes the maximisation of value for the minimum cost. Money is a convenient scale against which effort, sacrifice, comfort, danger, risk and numerous other factors may be measured.

Many authors have suggested that money is an objective measure of cost while satisfaction is a subjective or intangible measure. Markus[1], however believes that this distinction is not valid and money is no more objective than desire, satisfaction or sensation. As an example he looks at experiments of Mosteller[2], in which people's evaluation of the worth of the money itself is measured. The relationship is not only non-linear but varies from person to person and group to group. It is not even static for an individual. Under conditions of risk, danger or poverty, an individual's judgement of the value of a certain increment, or decrement of money will be quite different from his judgement in the absence of such conditions.

Cost Benefit Analysis

The study of the broad implications of several alternative designs with the aim of optimising, say cost, for the whole community has been given the name Cost Benefit Analysis. It is a practical way of both assessing the desirability of projects where it is important to take a long term view, and also allowing for the side effects. Unlike private investment appraisal, which is based solely upon the costs incurred and the likely returns, Cost Benefit Analysis sets out to place a value upon the losses or gains of the whole community, including many for which money cannot be exchanged. In the past, Cost Benefit Analysis has been used to provide a Yes/No answer to larger projects of national importance. It is unrealistic to expect projects financed privately to be the subject of a complete Cost Benefit Analysis, but many of the principles may still be usefully employed in the Private Sector.

The System Being Optimised

Organisations which need buildings to house their activities have certain definable objectives to meet. In order to achieve these objectives a pattern of behaviour and activity takes place. The setting and environment for this activity will influence the degree to which the organisation will succeed in meeting its objectives. Thus the designer is dealing with a complex and interactive system which has been represented by Markus[1] in Figure 1.

The objectives of the organisation may be social, economic or cultural whilst the activity pattern to meet these objectives may include activities directly related to

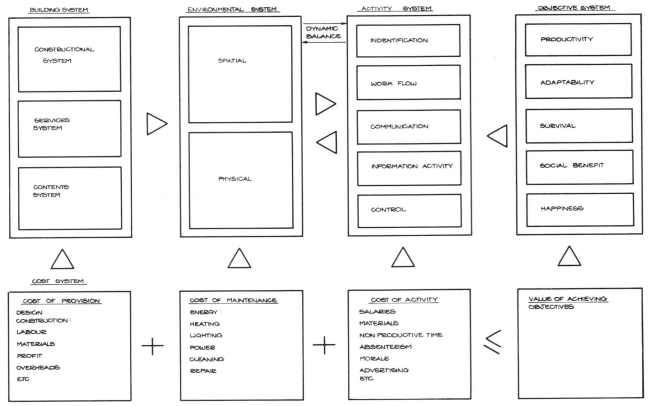

Fig. 1   *Conceptual model of the system of building and people (After Markus [1]).*

production, communication, control or identification. Depending on the nature of the objectives there will generally be a host of informal activities which are often vital to the survival of the organisation and its morale.

The environmental system contains two main elements, the 'spatial' and the 'physical'. The spatial element can be described and measured in terms of size, number and type of spaces and the inter-connections between them. The 'physical' element defines the 'spatial' element in terms of the visual, thermal, acoustic and other physical characteristics of the space.

A cost system is also attached to the conceptual model as each of the four parts has an associated cost or revenue. There is a cost for providing the hardware, maintaining the environment and providing the activity. The objectives also have a value which is partially or wholly met by the sum of the other components. It may be argued that the value of the objective when considering health, education or leisure activities, is given by the cost of providing the building, environmental and activity systems. In respect of commerce, the revenue from the objective needs to be greater than the cost of provision.

Optimum Cost

The optimum cost for a building system is that cost which is the least possible, over the life of the building, when all the user requirements and constraints have been satisfactorily met. In the building industry the constraints are more commonly referred to as the design criteria or the specifications, and are applied to each of the four systems shown in Figure 1. The optimum cost can also be defined as that which maximises the return on the client's investment. It should be noted that in many cases the return will not be measured in purely monetary terms.

If investment and benefit could be measured on identical scales, the optimisation procedures would be simple. Attempts (Markus[3]) have been made to cost the subjective aspects of our requirements from buildings, but as yet there are no fixed rules by which designers can incorporate subjective data within investment appraisals. This is

104

probably the first major difficulty encountered when trying to optimise the complete system. Another is the time scale involved. Even if all the user requirements are known at the outset of the design, they are unlikely to remain constant throughout the life of the building, perhaps some sixty years. The items of cost which form the cost analysis and the form of treatment undertaken will vary from one sector of the building industry to another. Differences arise partly from the differences in the responsibility clients have for the costs arising from their use of buildings and partly from the differences in the incidence of taxation, subsidies and grants.

Sub-Systems Optimisation

The building has already been described as consisting of five sub-systems but within each of these there are many more sub-divisions. Taking the environmental system, sub-divisions might include central heating and cooling production, the distribution networks and so forth. These sub-systems all interact with each other to differing degrees, in a complex way.

If it is possible to rank sub-systems into a single linear order of dependence, then optimising each sub-system in the reverse order should yield the optimum solution. This is the basis of *dynamic programming* and is used widely within the chemical engineering industry. The main criterion is that there should be no feed-back within the total system. Unfortunately the building system cannot be treated in this simple fashion, but it is possible to make an approximation to this idea by optimising sub-systems and then checking that the solution obtained is not strongly dependent on other sub-systems. If systems are strongly dependent, for instance, lighting level, window area and heating costs, they should be treated as one and optimised accordingly.

In order to be able to make a start at optimisation, it is necessary to leave the subjective aspects to one side and concentrate on objective measures. Many aspects associated with the engineering services plant are easily quantifiable and are not strongly dependent on subjective measures. Hence this is an area where optimisation can easily be performed.

ECONOMIC APPRAISAL TECHNIQUES FOR SUB-SYSTEM OPTIMISATION

Investment Appraisal

When cost is being optimised, investment appraisal can often be used to choose the best of various feasible solutions and point the way to the optimum solution.

There are two basic groups of investment appraisals; those which aim to compare two or more solutions, and those which measure the profitability of one solution. It may be quite possible to say that one solution of three gives the best return on capital invested, but investment in this project might not be economically justified as the rate of return does not reach a predetermined level. It is proposed to review some of the more common methods of investment appraisal.

Equivalent Annual Cost

The equivalent annual cost method of comparison takes account of both capital and recurrent expenditure over the full period of the study. An acceptable interest rate is chosen and all lump sum payments or receipts are amortized using the uniform series capital recovery factor:

$$\frac{(1 + i)^n \, i}{(1 + i)^n - 1} \quad \text{where} \quad \begin{aligned} i &= \text{interest rate} \\ n &= \text{number of years} \end{aligned}$$

(This and other factors are derived and tabulated by Pilcher[4]).

Annual payments present no difficulty and are added to the amortized capital to produce the equivalent annual cost. This procedure is carried out for each of the design solutions being considered and the one with the lowest equivalent annual cost is selected as being the nearest to the optimum.

105

Present Worth
Present Value

The present worth method, alternatively known as present value, is sometimes easier to apply than equivalent annual cost, especially when widely varying sums of money are paid out or received over a period of time. All future sums of money, both positive and negative, are converted to a common base at the beginning of the project. Again an interest rate is chosen which represents the value of the capital involved or an acceptable rate of return for the money. The factor used to bring all sums of money back to the common base is:

$$\frac{1}{(1 + i)^n} \quad \text{where n is the number of years from the base plate}$$

The equivalent annual cost as a method of comparison is more readily understood than present worth, as the figures produced tend to have more significance. Present worth calculations are easier to apply when the cash flows are irregular, but the total present worth for the project has little meaning except for comparison purposes. However, the use of either method should lead to the same conclusions.

Discounted
Cash Flow

The two previous methods assume an acceptable interest rate at the outset whereas the method of discounted cash flow is concerned with calculating the rate of return that a given set of cash flows is able to support. This is sometimes known as the internal rate of return or the DCF yield.

Once a complete set of cash flows, both positive and negative have been calculated they are brought to a common base using a present worth factor. An interest rate is guessed to be the DCF yield. The aim is to choose an interest rate which will cause the sum of all the discounted cash flows to amount to zero. If the sum does not equal zero, but is positive, a higher interest rate is chosen, and the calculation repeated. The converse applies for a negative sum. Interpolation is usually required between the interest rates which yield totals just above zero to obtain the exact rate of return.

If negative cash flows occur other than at the beginning of the project, a dual rate discounted cash flow yield calculation is required. At the outset a fixed rate of interest is chosen which represents the cost of money while the project is being financed (i.e. negative cash flow). The calculation will then provide the actual return after interest on borrowed money has been paid. The difficulty with these and other forms of investment appraisal is in establishing a set of cash flows for the project.

Break-Even Cost
Analysis

Break-even cost analysis is concerned with making comparisons between alternatives where the cost of each is affected by a single common variable. The points at which different solutions cost the same are known as the break-even point. The method is best illustrated by an example.

Example. It is necessary to install a condenser cooling water system to pump water from a refrigeration machine to the cooling towers. If a 150 mm pipe is used, the capital cost of the installation will be £4,000 and the estimated running cost of the pump is £0·621 per hour. If a 200 mm pipe is installed there will be a capital cost of £8,000 and a reduced running cost of £0·426 per hour. A 250 mm pipe will cost £12,000 and £0·291 per hour. The life of the installation is forecast as being 20 years after which there is no salvage value. An interest rate of 11% is assumed and the problem is to determine the most economical pipe size when 4,000 hours pumping per annum is expected.

Initially the capital cost of each installation is amortized over the project life using the given interest rate. The cost of pumping is then calculated and added to the amortized capital as shown in Table 1.

For a pumping period of 4,000 hours per annum the 250 mm pipe is most economical. If a break-even chart is constructed, as in Figure 2 it can be seen that if less than 2,648 hours pumping per annum is envisaged the 150 mm pipe will be the most economic. Between 2,648 and 3,386 hours the 200 mm pipe would be a better proposition. The main difficulty with this analysis is in the estimation of the running costs over the life of the installation.

Fig. 2

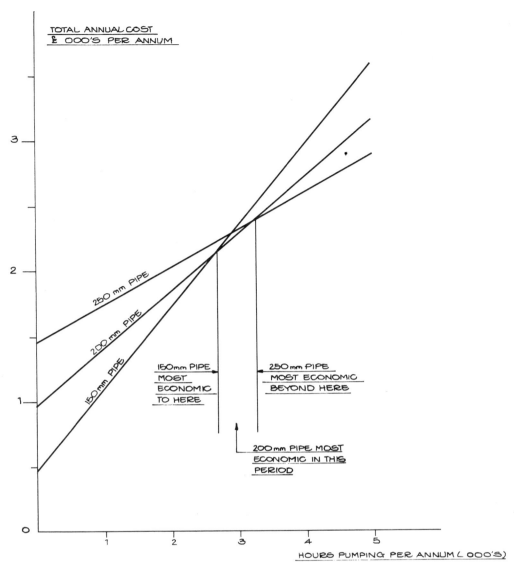

Break-even chart for pumping example.

Table 1

Break Even Cost Analysis

Diameter of Pipe	150 mm	200 mm	250 mm
Capital Recovery on initial cost of pipe	$4,000 \times 0{\cdot}12257$ = £490	$8,000 \times 0{\cdot}12257$ = £981	$12,000 \times 0{\cdot}12257$ = £1,471
Cost of 4,000 hours pumping	$4,000 \times 0{\cdot}621$ = £2,484	$4,000 \times 0{\cdot}436$ = £1,744	$4,000 \times 0{\cdot}291$ = £1,164
Total Annual Cost	£2,974	£2,725	£2,635

The Capital Recovery factor is:

$$\frac{(1 + 0{\cdot}11)^{20} \times 0{\cdot}11}{(1 + 0{\cdot}11)^{20} - 1} = 0{\cdot}12257$$

107

The Optimum Thickness for Thermal Insulation

It is well known that the heat lost from a pipe may be reduced by applying greater thicknesses of insulation, but the capital cost is increased. Depending upon the values assigned to a large number of variables there will be a minimum total cost, and a corresponding insulation thickness.

The methods proposed in the relevant British Standards and the Codes of Practice for calculating insulation thickness do not take into account the time value of money. If money is to be invested in additional insulation then some measure of the value of money must be taken into account, possibly with an assumed inflation rate.

In most cases designers will not calculate economic thickness, but resort to specifying the thicknesses tabulated in BS 1334[5]. These, however, are minimum values to meet the standard and are not suggested as the values to be specified in every case. With the present high cost of energy there is a strong case for calculating the economic insulation thickness.

A computer programme written by the author does the necessary arithmetic and outputs the results as shown in Table 2, together with the input information. The optimum thickness is represented graphically in Figure 3. (This graph can be applied to many situations where an increased capital expenditure produces a reduced running cost).

Table 2

Optimum Insulation Thickness by Tabulation

Pipe—89·0 mm O/D   **Black Steel to BS 1387:1967**   **Service:LPHW**

Insulation Thickness (mm)	Insulation O D (mm)	Insulation Cost (£/m Run)	Total EAC* (£/m Run)	Heat Loss (W/m Run)	Insulation Surface Temp°C	Cost of Heat Loss (£/m Run)	Total Cost (£/m Run)
0·0	89·0	—	—	210·0	—	72·69	72·69
12·5	114·0	1·41	2·30	50·0	32·7	17·32	19·62
19·0	127·0	1·53	2·49	38·0	28·7	13·14	15·63
25·0	139·0	1·65	2·69	31·6	26·6	10·93	13·62
32·0	153·0	1·96	3·19	26·8	25·1	9·26	12·45
38·0	165·0	2·20	3·59	23·9	24·2	8·26	11·85
44·0	177·0	2·58	4·21	21·7	23·5	7·51	11·72
50·0	189·0	2·90	4·72	20·0	23·1	6·91	*11·60*
63·0	215·0	3·82	6·22	17·3	22·3	5·98	12·20
75·0	239·0	4·44	7·24	15·6	21·9	5·38	12·62

INPUT DATA

1. Mean surface temperature of pipe (77·5°c)
2. Average air temperature (20°c)
3. Thermal conductivity of insulating material (0·044 W/m°c)
4. Surface convection coefficient (11·0 W/m²°c)
5. Evaluation Period Years and Hours per year (10 year and 6048 hours per year)
6. Cost of energy (0·00159 £/MJ)
7. Overall diameter of pipe (89·0 mm)
8. The heat loss from an uninsulated pipe (for camparison purposes) (210·0 W/m run)
9. Insulation cost (as in Table 2)
10. Interest rate (10%)
11. Inflation rate (0%)

The figures in brackets refer to the data used in compiling the table

*Equivalent Annual Cost

Columns 1 and 3 form part of the input data whilst column 2 is calculated by adding twice the insulation thickness to the overall pipe diameter. Column 4 gives the total equivalent annual cost for each thickness of insulation. This is calculated using the uniform series capital recovery factor, previously mentioned, multiplied by the number of years in the evaluation period, to give the total equivalent cost. Column 5 gives the heat loss per metre run and is calculated using equation C3.10 from the IHVE Guide, Book C[6]. Column 6 gives the calculated surface temperature of the insulation using equation C3.11 again from the IHVE Guide. The cost of heat loss, over the evaluation period is calculated from the following equation:

$$\text{Cost of heat lost} = 3,600 \times H \times q \times Y \times 10^{-6}$$
$$\text{Where } H = \text{evaluation period in hours}$$
$$Q = \text{heat lost W/metre run}$$
$$Y = \text{cost of heat £/useful MJ}$$

The total cost, column 8, is the summation of Columns 7 and 4. It can be seen that there is a minimum value corresponding to an insulation thickness of 50 mm.

Fig. 3

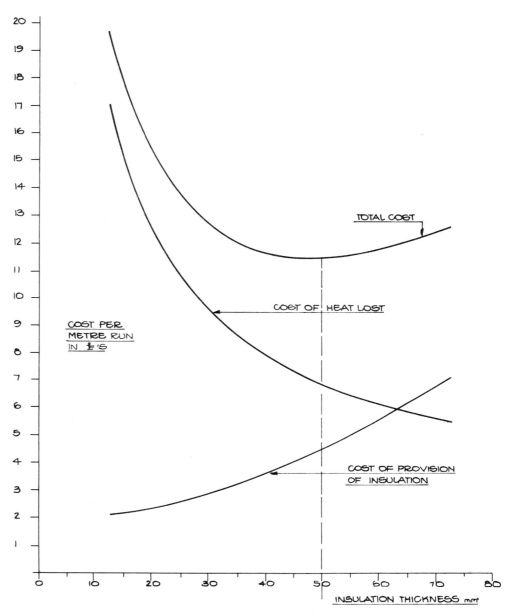

Cost per metre run for an 80 mm nominal bore pipe carrying LPHW, plotted against insulation thickness.

If inflation is to be taken into account, it is assumed to occur at a constant rate, throughout the evaluation period, and is simply a case of compounding the energy cost at the chosen rate.

For general use of the programme, only the data relating to the optimum insulation thickness for each pipe size, need be printed. By altering the values of the input data it is possible to use the programme for all piped services. The main differences will be due to the different surface temperatures of the pipes and the expected period of operation. When special insulations are required, for instance to achieve a vapour barrier, the cost of sealing should be included in the insulation cost.

The results for several interest rates and inflation rates and a range of pipe sizes are shown in Table 3. It is found that there is little variation in the optimum thickness for varying interest and inflation rates. The difference is more marked however in the case of hot water service pipework, due to the larger number of hours operation per year. 44 mm thick insulation is not generally available in pre-formed rigid glass-fibre sections for pipes less than 80 mm nominal bore. The results however tend to suggest that 44 mm would be an economic thickness for 40, 50 and 65 mm pipes if it were available as a standard thickness.

The logic used in the programme is simple and straight forward and lends itself to the larger desk top calculators. If these machines were available in the design office, with the cost data periodically updated, it would be possible for the designer to calculate the economic thickness for each individual set of circumstances within a very short space of time.

Table 3 **Calculated Economic Thickness for Low Pressure Hot Water Piping**

Interest Rate %	10	14	14	14	14	14
Inflation Rate	0	0	5	10	15	20
Pipe N/B* mm						
15	32	32	32	38	38	38
20	38	32	38	38	38	38
25	38	38	38	38	38	50
32	38	38	38	38	50	50
40	38	38	38	38	50	50
50	38	38	38	50	50	50
65	50	38	50	50	50	75
80	50	44	50	50	63	75
100	50	50	50	50	75	75
125	50	50	50	50	75	75

Nominal Bore

COST MODELLING

The Cost Model

A cost model is a mathematical statement of cost for a set of projects with a common set of identifiable variables. The cost of a primary school might be expressed in the form:

$$Cost = A + Bx_1 + Cx_2 + Dx_3 + \ldots etc,$$

Where A, B, C . . . are constants determined by the type of schools considered, and x_1 to x_n are the variables which will describe each particular school. For instance the following might be significant variables:

x_1 — total number of children
x_2 — number of classrooms
x_3 — area of the hall
x_4 — number of storeys
x_5 — area of circulation space
and so forth

Regression Analysis

If such a model was available, the designer could enter the values of x_1 to x_n for his design and quickly calculate the total cost. Attached to the model would be an assessed accuracy so that it is possible to assess the reliability of the cost produced by the model.

Regression Analysis is a statistical technique which will produce an expression, or mathematical model, from sets of observed data. It must be borne in mind that the technique will always fit a model to the data and statistical significance tests should be applied to measure the 'goodness of fit'. This will become apparent later.

Returning to the previous example, it is desired to model the cost of school buildings using one variable. It might be decided that due to a similar standard being maintained for all the schools in the data sample, the cost will be most dependent on the number of children within each school. Having collected a number of sets of data these can be plotted as points, as shown in Figure 4. The method of regression analysis aims to choose the regression line, or line of 'best fit', by minimising the sum of all residuals. These are the differences between the observed value of cost and the cost given by the regression line. The differences are squared to eliminate problems with positive and negative differences cancelling each other, in part or completely.

Fig 4

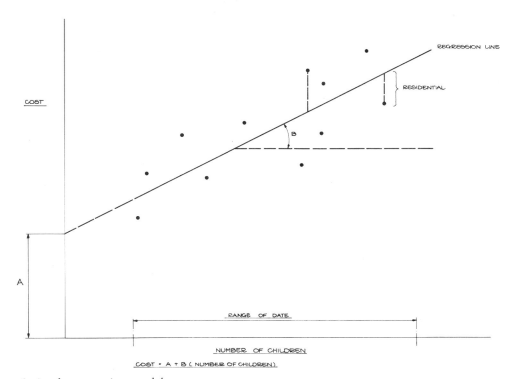

A simple regression model.

A correlation coefficient, or measure of 'goodness of fit' can be calculated as a function of the sum of the residuals in both the x and y directions. If a correlation of 100% is obtained, all the residuals will be zero and the regression line will pass through every point.

Having established this simple model, it is possible, knowing the number of children, to predict the cost of any proposed school. Two points should be noted here. The model is only suitable for predicting the cost of a new building similar in type to those used in constructing the model. Secondly, prediction outside the range of observed data should be undertaken with care. The assumption of linearity may be justified over the range of observations available but there is no evidence that the relation is linear beyond this range.

If a model was constructed as outlined above, it would have a poor correlation coefficient. This is because the single variable chosen does not explain all the cost variance. In order to improve the model more variables need to be introduced. Two variables can be represented geometrically as a plane in space but beyond this, graphical representation is not possible.

There is no limit to the number of variables which can be included in what is termed a multiple regression analysis. Initially more variables will be included than are necessary and then once the model is constructed each variable is validated for the inclusion in the final model. The aim is to calculate the proportion of variance in cost explained by each variable and see if this portion is statistically significant. Insignificant variables are then discarded. This process is known as backward elimination and will show which of the variables initially chosen for inclusion are not independent. Other methods of selecting the best regression equation invole forward selection tests. As more variables are included in the model, so the number of data sets required increase. McCaffer[7] suggests that as a rough guide the minimum number of past schemes required is two or three times the number of variables included in the final model.

The technique so far described, is based on items of data which are assumed to vary linearly with cost. This limitation can be overcome by transforming the data before applying the model. For example, if the cost of the school gradually tails off with the increasing area of the hall, then this variable in each set of data could be raised to a power to linearise the relationship. For a full statistical analysis, reference to Draper and Smith[8] is recommended.

To construct a comprehensive cost model, access is required to sufficient sets of data from past schemes, and a certain amount of processing is usually required to extract the required information from the drawings and specification. Carrying out the steps described for a large volume of collected data will require the use of computers. There are several computer packages available which will do all the necessary calculations automatically, but care is required when interpreting the results.

Examples of a cost model

McCaffer[7] describes a number of cost models. One in particular Gould[9] deals with the capital cost of heating and cooling services in a wide range of buildings. He uses seven categories of air treatment from local space heating without fresh air to full air conditioning with centralized plant. In order to analyse the services distribution within each of a wide range of buildings, which includes a church, swimming baths and housing, each building is broken down into a number of sub-volumes. Data is then collected from each sub-volume. The variables used in the final model include functions which describe the heat and air flow through the building, the energy source and its distance from the terminal units. The environmental differences across the boundaries of each sub-volume and data concerning percentage of fresh air, heat gain, etc, are also included. The model produced will predict cost within a plus or minus 10% band for projects with a capital cost in excess of £10,000. This is probably more accurate than conventional estimating accuracies.

Optimising the model

This is not as easy as it would appear at first sight. Firstly, as the model produced is essentially linear, simple differentiation processes are precluded. Secondly, the model as a whole predicts costs within its accuracy limits and individual variables cannot be altered in isolation. Variables within negative coefficients cannot be increased to reduce the overall cost without reference to the practical situation. Some variables will have negative coefficients to compensate for other variables explaining more than their share of the cost variance. Therefore the model must be used as a whole.

Optimisation is possible, but only by trial and error methods at present. Experienced designers should have a 'feel' for the solution with the minimum cost and the model can be used as a rapid check. Initially, a set of values are chosen for entering into the model which look suitable and within minutes the corresponding cost can be obtained. These can then be changed, still maintaining a feasible solution, and a new cost estimate prepared. The iterative process can be repeated as many times as necessary to produce the minimum cost while still meeting the criteria.

If the design criteria are presented in a suitable form it should be possible to programme a computer to perform the necessary iterative process, checking at each stage that the design criteria have been met. It must be stressed that this procedure will not design a building, but will only enable the broad design strategy to be checked. Detailed design must follow in the normal manner. It should be possible however to keep checking the design as it progresses against the cost model to ensure that the optimum is being maintained.

CONCLUSIONS

Optimisation of a building design is a very complex process with absolute success difficult to attain. Good design solutions can only be measured against existing buildings and as such, optimisation becomes a continuous process, learning from the failure or success of previous designs. The main message of this chapter is to take as broad a view as possible of the implications of each design decision. The limiting factors will be the time available within the design period and the design team's general attitude to the long term consequences of each decision.

Time should be allowed within the design period for a critical appraisal at regular intervals. In order that the results of these appraisals are to be of any use, there must be the facility for going back at least as far as the previous appraisal, and solving the problem in a more efficient way. If appraisal is carried out at frequent intervals, the redesign of particular aspects should not present too much difficulty. If, however, the design is almost complete before an overall view is taken regarding capital and running costs and their relative balance, there is little chance of carrying out the fundamental changes that may be required.

The process of appraisal should not cease when the design is complete, or even when the construction period has ended, but should continue throughout the life of the building albeit at increasing time intervals. This appraisal is concerned with two basic areas, firstly, the suitability of the design to the user's requirements and secondly, the cost of running and maintaining the building and its services.

Sub-system optimisation requires a great deal of time developing the techniques initially but if the work done for one particular project is carried out in a generalised way, it should be possible to utilise the experience gained in connection with other projects. This applies particularly to the development of computer programmes, where the development costs would be prohibitively high if charged to one project but if spread amongst a large number of projects, the computer can save a great deal of valuable design time. The programme developed for pipe insulation is a good example. Once written it may be re-run with as many different sets of input data as required.

Total optimisation should be a way of thinking rather than a definite method to be followed in all cases. It is hoped that the techniques described will assist in the decision making processes within the engineering services and other sectors of building. The science of decision making is growing rapidly and designers should keep abreast with developments in this field. Even more important is the need to relate these developments to the specific problems found within the construction industry.

Acknowledgement   This paper has been based on a project undertaken at Loughborough University by the author. Thanks are due to the academic staff for their guidance.

REFERENCES

1   Markus, T. et al. *"Building Performance."* Applied Science Publishers (1972).
2   Mosteller and Nogee. *"An Experimental Measurement of Utility (1951)."* Edwards and Tzersky (Editors), Selected Readings. Penguin Books (1967).
3   Markus, T. *"The Real Cost of a Window—An Exercise in Cost Benefit Study in Building Design."* Building Performance Research Unit, Strathclyde University (1967).
4   Pilcher, R. *"Principles of Construction Management for Engineers and Managers."* Mc Graw Hill (1966).

5   BS 1334:1966. The use of Thermal Insulation Materials for Central Heating and Hot Water and Cold Water Supply Installations.

6   IHVE Guide. The Chartered Institution of Building Services. London (1970).

7   McCaffer R. *"Some Examples of the use of Regression Analysis as an Estimating Tool."* The Quantity Surveyor, December 1975, Volume 32, No. 5.

9   Gould, P. R. *"The Development of a Cost Model for H & V and Air Conditioning Installations in Buildings."* MSc Project Report. University of Loughborough (1970).

8   Draper and Smith. *"Applied Regression Analysis."* Wiley Interscience (1966).

DISCUSSION

**Dr M. F. Green
(Department of Health and
Social Security)**

I would like to make a number of points relating to cost optimisation methods and cost modelling techniques. I do not think Mr Carver has adequately highlighted the limitations of such methods. He stated at the outset that no attempt had been made to evaluate subjective factors in the cost optimising exercises he has presented. However, I believe that costing these factors is crucial if cost optimising techniques are to be used in making decisions. I suggest that if all factors are not included in the model then only if order of magnitude differences are found to exist between A and B can we confidently base a decision on the results of an optimisation exercise. However, if order of magnitude 'cost' differences exist between A and B then the designer could have probably accurately predicted subjectively that A was better than B.

My second point relates to cost modelling. I do not think it is as simple as Mr Carver suggests. The concept that one can acquire a mass of data which when processed by a computer using a multivariable regression program produces a meaningful result is misleading. I believe it is much more difficult than this. In the first place data is very difficult to obtain and secondly I have rarely seen a regression model where significant practical benefits have resulted from using more than 3 or 4 variables. Mr Carver's example of costing schools highlighted one important mistake when trying to identify variables. He only suggested three possible variables, and yet two of these variables, namely the number of classrooms and the number of children are both functions of one factor—school size. Although a regression program would inevitably produce a result, just as accurate a predictive model could probably have been obtained by simply regressing school cost against the area of teaching space.

My final point concerns the illustration of the linear regression technique. I did not see a confidence interval on this graph. I believe it is extremely important when discussing regression analysis, even when simply in descriptive terms, to include the concept of confidence intervals, thus ensuring that the predictive powers of regression techniques are portrayed in perspective.

M. N. Carver

Admittedly in the paper I did leave aside subjective aspects but I hope I did not give the impression that they should be left out of any subsequent analysis. In building design there are many areas where quantifiable measures are not easily taken so I concerned myself with the mechanical service where it is possible to put some figures to what, hopefully, are facts.

I did mention in my paper the need to look at the sensitivity of the optimisation process, for instance, in connection with the insulation thickness example. I pointed out that there were quite a large number of insulation thicknesses that did not show very much cost variation. Common sense should prevail and not an over indulgence in refined methods where they are not needed. The availability of capital in the first place is obviously a primary consideration.

Regression analysis is a very difficult and time consuming process, possibly only taking place in the research context. I would suggest that gathering the required data requires one day per contract. A quantity surveyor would be required to go through the drawings, specifications and all the cost data for that contract, extracting the useful information. Forty or fifty sets of data may be needed to construct a reasonably accurate model, there is a large amount of work to be done and this is only for one model for a specific type of building. At the national level it is possible for Government to do this form of modelling for schools, hospitals, offices and other types of building but generally not feasible for the small private office. The regression line in Fig 4 without confidence limits was constructed purely for demonstration purposes. No data was used and no scales were placed on the axes. If a full regression analysis is carried out and all the statistical techniques are applied, very accurate confidence limits can be achieved and an estimate of the accuracy of the model produced with a known certainty.

**T. Smith (Steensen
Varming Mulcahy &
Partners)**

Harry Truman, the draper from Missouri who became President of the United States, once said that in dealing with the problems of the present and the future the answers could always be found in history. I subscribe very much to that opinion and I was reminded whilst listening to Michael Carver give his paper, of an event that occurred about 15 years ago. Until his recent retirement I had the privilege of being in partnership with an eminent engineer called Varming who is about 20 years my senior. I have the reverse age relationship with Michael Carver. All those years ago Varming said to me 'you know, I do not understand you young fellows with your computers and your mathematical approaches to the job. It does not bother me that I do not understand it, what does bother me is that I am not bothered.' It was said facetiously I know, and in the same vein I now say that I do not fully understand what Michael Carver is describing in his paper.

I would, however, very much like to hear his opinion on how these theories can be applied in a practical sense within the design office, keeping in mind that the design office consists of people of varying intelligence and education, it will be essential that the development of the techniques described results in data being available which are easily transcribable and directly applicable to the day to day work of the design office. How does he see this development occurring?

H. G. Mitchell (H. G. Mitchell & Partners)

Regarding the transition from mathematics and computers to the design office, I see this transition as a very slow process. There is still a great deal of research to be done into devising methods and probably still more research in making these available to the designer within the design office. As Mr Smith said earlier (p 100), time is the all-important factor. Is there time to do this work while one is designing the building at the same time, or should they be done within research institutions and then presented to a design office in a form that is readily used. There is then a danger, if methods are presented without the design office knowing the full theory behind them, that they will be applied blindly without true feeling for the results that are produced.

Professor P. Burberry (University of Manchester)

The application of optimisation techniques to building design is very difficult. Situations where one function dominates all the others and where the building must be designed in accordance with that function, lend themselves well to optimisation of the specific function. However, in most building design situations, many different aspects are involved and it is very difficult to establish any systematic relationship between them. Dealing with each one separately is not of much assistance since the individual optimum solutions for each would almost certainly be in conflict and there is no rational way of establishing a balance.

In practice, it is questionable whether optimisation is a fruitful means of approach. Normally many constraints are attached to the design of a building and, quite soon in the design, a limited number of possibilities begin to emerge. Town planning constraints may govern form and height of buildings, functional requirements will determine plans and daylighting needs may govern building dates and fenestration. If a few possible solutions emerge from the designers evaluation of these constraints, then the important need is to have quick methods of evaluating the possibilities rather then establishing optima, which are quite likely to be outside the range of possibility and which do not tell the designer which of the possible solutions is best.

R. J. Wilde (Westminster City Council)

I would like to refer to the disturbance factor in cost benefit analysis. I envy my counterparts in industry who are able to evaluate the cost of plant breakdown time to very close limits of accuracy. In buildings, costs can be evaluated of plant maintenance and of standby but how to evaluate downtime or replacements or the extent to which standby services should be provided, is quite a different matter. For example, how does one allow—at the design stage—for the removal for six weeks and consequent loss of service to tenants, of the one lift in a medium-sized block. From a heating point of view it is fairly easy. Heating charges could be reimbursed but the problem is the abstract indefinite items, like the loss to the community for a week of a library, day centre, etc.

M. Carver

Cost benefit analysis in the past seems to have been applied to large projects of national importance where a 'Yes' or 'No' answer was required. In these analyses a certain figure was put on the value of a life and the cost of amenity. These must be estimates in the view of Government or those carrying out the analysis. A major problem in using cost benefit analysis in the building context, particularly, with reference to the downtime of services plant is the kind of value placed on staff. Obviously, if they are very busy and productive it is a very serious problem when their transport throughout the building is hindered. Perhaps it might just mean a slight overcrowding in the next lift or, possibly, a long wait. Evaluating the worth of people is likely to be a highly controversial subject but until figures based on consensus of opinion are available it will be very difficult to apply subjective measures within optimisation procedures.

Chapter 9

Cost Implications of Materials Structure & Form

P. Burberry

Environmental control is not merely an ancillary aspect of design, but, in most climates, one of the basic functional reasons why the building is required. At present we are entering a new era of possibility in the relationship between building form and environmental control. For the whole of recorded history buildings have been governed in their size and form by the need for natural lighting and natural ventilation. The need has been so totally integrated into virtually every existing building and into the process of design that, except for large city blocks where light wells had to be provided, few architects have been conscious of the limitations within which they were, in fact, working. All design thinking took place within the constraints imposed by natural ventilation and lighting and designers were only conscious of decisions involving aspects such as structure and planning. Heating also imposed a number of similar constraints upon design.

The situation has now completely changed. The fluorescent tube provides an acceptable form of lighting at relatively low cost and without the unacceptable heat loads resulting from previous lamps. Electric fans, and the associated apparatus which has developed for filtering and conditioning air mean that ventilation can be achieved and controlled without needs for windows and in spaces far larger than could have been controlled by window opening. Central heating systems and refrigeration plant enable heat to be input and extracted from rooms without the need for direct linkage with the outside environment.

Designers are faced with a new decision problem; whether or not to accept mechanical control of environment. The decision should be governed by considerations of cost, psychology and physiology. Medical opinion varies as much as architectural fashion. Not long ago, it would have been thought that an entirely internal existence would be unhealthy and that sunshine and fresh air were essential to well being. This is no longer the case. The value of sunshine is apparently not significant with present day diets and adequate fresh air can be provided by mechanical means.

The psychological problem is very different, it appears that an awareness of the outside world is something that is very much needed by many people and that total isolation from the external world should not be done without very careful thought. In the recent era of cheap energy many people argued that, even though the buildings required continuous use of electric light and fan power and cooling for much of their life, the large building volumes made possible by mechanical environmental control are economical in heat losses, of materials of construction and also permit dense occupancy of land with consequent economy in communications, roads and services.

The present energy crisis requires a re-evaluation of this balance. The lighting and cooling of large buildings uses electricity which, because of the large amount of primary energy required to produce it, is becoming rapidly more expensive and well designed buildings, taking advantage of the natural environment and using other types of fuel only for space heating seems very sensible in present day terms.

The dramatic differences in the form of buildings which can result from the change between natural and mechanical control of environment are compared in Fig. 1 and Fig. 2. Fig. 1 shows a simplified view of the original (1932) Sir Owen Williams designed, Boots factory at Beeston, Nottingham. It has a large production area of 68,7000 m²: 40×10^4 m³ (740,000 ft² : 14×10^6 ft³) and to achieve functional juxtaposition of storage, manufacture and packing and provide adequate light and ventilation a very complex form was devised which gave a very satisfactory level of environmental performance. The diagram shows the essential elements of the form from the environmental viewpoint. The factory for Players shown in Fig. 2 and designed by Arup Associates, (1971), is sited only a short distance from the Boots factory although its vital statistics of: 57,700 m² : $41 \cdot 8 \times 10^4$ m³ (621,000 ft² : $14 \cdot 75 \times 10^6$ ft³) are essentially the same as the Boots factory, the buildings are very different. The Players factory has a need for humidity control but basically it provides, as does the Boots building a flexible area for storage, manufacture and packing of small, light items requiring a clean environment. Its shape is made possible because of the mechanical control of environment.

It is interesting to see that the very large volumes enclosed in light wells in the Boots building are matched by the volumes occupied by services in the Players factory and mechanical control has not enabled volume to be reduced.

A similar change in form took place in office buildings which involved mechanical control of the environment although in this case the change of form gave rise to the need for mechanical control. It is fundamental and salutary in this connection to note the misconceptions about the performance of high buildings which were entertained by their advocates even though very simple investigation would have brought to light the realities of the situation. In the 1930s many European architects advocated high

Fig. 1

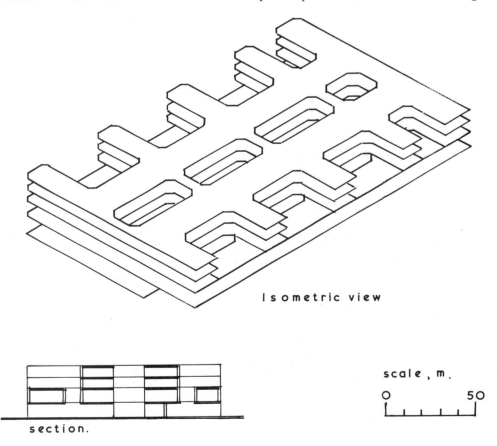

I s o m e t r i c   v i e w

section.

scale , m.

0   50

Boots Factory, Beeston, Nottingham 1932. Simplified isometric view and section. The isometric shows clearly the complex form devised to allow natural light and ventilation and to allow the processes of storage, manufacturing, packaging and despatch to operate smoothly. (Compare with Fig. 2 to same scale).

Fig. 2

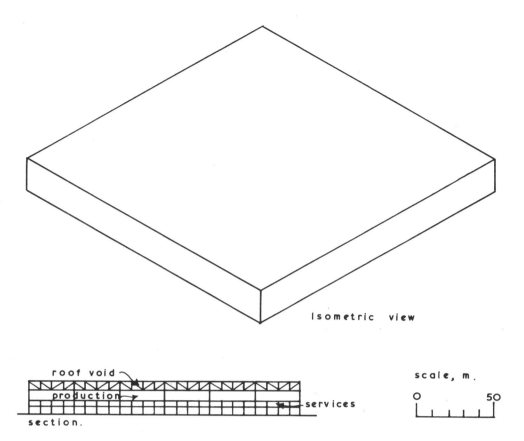

Isometric view

roof void

production → services

section.

scale, m.

0   50

Players Factory, Beeston, Nottingham 1971. Simplified isometric view and section. Simple form with large, flexible production area on one floor. Note the roof void which contains 27 air handling plants and the complete floor devoted to services. (Compare with Fig. 1 to same scale).

buildings for their environmental advantages. In the cities of the time, although buildings did not rise much above 30 m, they formed continuous facades on either side of relatively narrow streets. The rooms facing the streets, particularly at low level, were somewhat dark, subject to traffic noise and, smell from the streets.

It seemed obvious to many architects that taller more widely spaced buildings would solve these problems and give better light and freedom from noise and fumes. Town planners and legislators also held these views and new rules were framed governing the form of buildings in central areas to permit high rise development. Whatever virtue resulted from this, the environmental results were disappointing. Free from all surrounding buildings and trees sky glare and solar gain created acute problems. Although the noise of traffic at the foot of the building became less at higher levels, the noise from more distant streets now reached the upper parts and noise levels remained high to the top. The wind and stack effects which act upon high buildings prevented the use of natural ventilation and this objective also was lost. In addition the velocities of air movement at ground level were intensified around very large buildings and open areas at ground level could become untenable even in gentle breeze conditions.

All these problems could have been predicted with the technological knowledge available or observed by visiting American high buildings. Designers of buildings are too inclined to use experience as a guide even in circumstances where no relevant experience exists rather than make technological predictions to aid judgement. It is also clear that, having achieved a preconception of the desired form both designers and all others responsible for decisions about building form were reluctant to place their concepts in jeopardy by further analysis.

The examples used so far are modern ones and deal mainly with building form in relation to light and air movement. It is desirable to consider thermal performance, which unlike light and air movement is powerfully affected not only by form and fenestration but also by the materials making up the fabric of the building. Buildings in many parts of the world have, long before the phenomenon involved were explained, achieved by trial and error and progressive improvement a remarkable adaptation of form and materials to give the best possible thermal comfort. The English thatched cottage, the North African Courtyard House, and the typical house for the hot humid tropics are very different from each other (Fig. 3) but each represents a very sophisticated solution to the problem of providing thermal comfort with economy of resource.

The English cottage is covered by thatch giving an excellent standard of insulation for most of the exposed surface. Wall and window areas are small. A central fireplace and chimney enable a single fire to contribute heat to the whole building. For an era when hand gathered peat and timber formed the only fuels it made very economical use of energy. The North African Courtyard has, of course, white walls to reflect some of the sun's heat, small windows and thick walls to delay the time at which the heat of the sun will reach the interior. The main significance of the design is, however the way in which the house takes advantage of radiation to the clear night sky to form a pool of cool air in the courtyard, by a process similar to that in a frost hollow. This air penetrates into the ground floor rooms giving increased comfort there. It is not affected by the suns rays until well on in the day when the sun is overhead. In addition a fountain in the courtyard gives evaporative cooling (the air is cooled by contributing heat to evaporate the water sprayed by the fountain).

The raised building from the hot humid tropics provides a solution to a more difficult problem. The main way of improving thermal comfort in these conditions is to increase air movement and the louvred walls, the raised floor and the balconies giving open-air sitting all contribute to this. In addition the low thermal capacity materials do not retain heat for long and so allow the structure to cool so that advantage can be taken of the somewhat reduced night time temperature.

It is worth dealing with these indigenous house types because of their direct relevance to the relationships between form, materials and environmental control and the way in which they demonstrate how comfort may be achieved not by the use of additional resource but simply by proper selection from available materials and forms. Beyond this, however, these houses have an important moral for designers. The explanation, in terms of physics and engineering, of the thermal comfort features of the design was far beyond the understanding of the designers but they were able to design at a level of performance beyond that of the science of their time, not using elaborate equipment, but exploiting the possibilities of the elements at their disposal, in particular the materials and form of buildings.

The present day situation is in distressing contrast. Present scientific knowledge enables specific understanding of most of the environmental phenomena in buildings and quantitative estimates are possible to a level far beyond that which is usually taken into account in present day design. Why is it that building, which used to be in the fore-front of applied science has lost its skill?

One interesting but isolated exception to this situation is an extension to St George's School, Wallasey, Cheshire, which was designed not to require a heating installation. The principles involved were well appreciated but, at the time of the design, impossible to quantify. The designer of the school, nevertheless had the courage of his convictions and produced a building, a section through which is shown in Fig. 4, which most people, and perhaps the designer himself, regard as being heated by the sun, but which in fact depends on the heat from the occupants and the lighting to a considerable degree. In the building the selection of materials and positioning of the insulation are intended to

Fig. 3

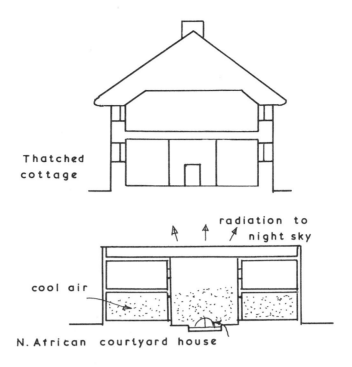

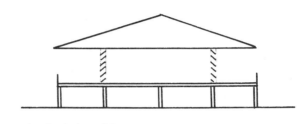

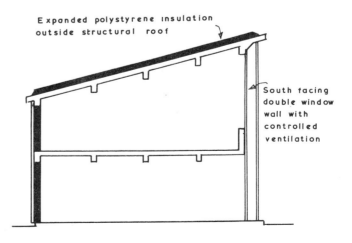

Indigenous house types which by their selection of form and structural materials achieve high standards of thermal performance.

Fig. 4

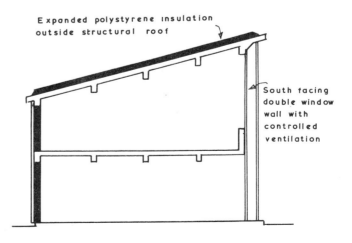

Simplified typical section through extension to St. George's School, Wallasey, showing massive construction and position and position of insulation and double window wall.

give a building which will absorb heat when the temperature rises, and when the sun is shining and retransmit it to the interior when the inside temperature drops. Ventilation is closely and ingeniously controlled to reduce heat loss when desirable and to increase it in periods of overheating. For a number of years the building performed to the satisfaction of its occupants without a separate heating system. It is not a method which can be widely used since comfort temperatures in schools are lower than in buildings used by adults and the density of occupation, and consequential heat gain from occupants, are very much higher than in other building types.

The building does, however, demonstrate that whether or not the heating installation can be totally dispensed with the form, and particularly the selection of materials and position of insulation, have a dramatic effect on thermal performance, and that, even in a scientific age it is still possible for designers to go beyond the limits of established design analysis.

Installations for collecting solar energy for space heating in buildings claim a great deal of current attention even though the economics of such installations cannot be regarded as satisfactory. Little thought is given by designers to orientation and fenestration in relation to utilisation of solar energy in the form and fabric of the building itself. The nett energy balance varies greatly with the orientation of windows and judicious planning taking this into account could give useful energy saving without any additional expenditure—in effect an infinite cost benefit. This is rarely done. Xenophon writing 2,000 years ago gives a more perceptive description of the principles of design to control and utilise sunshine than appears possible at the present. *'In houses with a south aspect the suns rays penetrate into the Porticoes in winter but in summer the path of the sun is right above our heads and above the roofs so there is shade.'*

Apart from the passing reference in connection with high buildings, little has been said so far about the interrelation of noise with form, and construction. While the forms of auditoria are generated by acoustic considerations and buildings are planned to separate noisy and quiet activities the major consideration in the present context is that of mass. Heavy walls give better sound insulation than lightweight ones. In recent years there have been more and more lightweight floors and partitions. Some prefabricated systems have abandoned even the masonry party wall between dwellings and replaced it with layers of plasterboard and very light internal partition in dwellings are now common place. The motivation for this is economy if materials and economy resulting from speed of erection. In this case, therefore, unlike those where overall building form is involved, and it is very difficult to isolate environmental cost implications from those arising from other aspects of the design problem, we see a clear situation where cutting cost has given rise to less satisfactory environmental conditions.

Unfortunately noise penetration does not manifest itself very directly at the drawing board stage and many architects have given it less weight than they would had they been able to experience, in advance, the actual conditions resulting from the design. In dwellings minimum standards are laid down, but in both dwellings and other buildings a judgement has to be made between cheap partitions, effective in every way except sound insulation and more expensive but more effective construction. It is a difficult decision and one of the few environmental ones where the issue is so clear.

The use of lightweight construction together with large windows had a marked but unanticipated effect on school environment. Twenty-five years ago it was demonstrated that economy of fuel in winter in schools could be achieved by lightweight structures with fast responding heating systems and this approval was widely adopted. The large windows intended to give adequate daylighting, but in fact, usually very much larger than necessary, because of the dictates of fashion, allowed the sun's rays to penetrate the lightweight, low thermal capacity structures which warmed up rapidly and the result was widespread overheating in summer—a consequence that not only was not considered at the time but which persisted in design for twenty-five years before recommendations emerged from the DES to limit overheating.

The picture which emerges from the whole consideration is one where, although the vital importance of building form and structural materials and finishes in governing environmental conditions is very clear, designers, in recent times, have not in general been motivated to give much weight to this aspect of design. The effect upon noise penetration and summer overheating has been disastrous. In the heat loss, light and ventilation fields cheap energy meant that there was no economic reason why building form should be carefully considered since electrical and mechanical installations would be provided to ensure habitability. The situation has changed and more careful designers consideration must be given to the basic fabric of the building.

Fig. 5

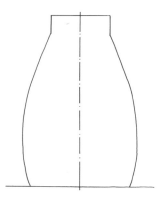

Elevation of high rise office building designed to give optimum evaluation. (Moseley 1963).

There are at present few well established design concepts related to environmental performance which are appropriate to modern buildings. A number of theoretical studies have been made almost exclusively in individual aspects of the problem. The ideal form to give maximum econmmy for one particular need is seldom a practical solution for a building. At the simplest level buildings of spherical form would have 20% less heat loss than the same volume in a cube. At a more sophisticated level studies by Lynn Moseley in 1963 demonstrated that the ideal form for a high building for a particular path of circulation would be that shown in Fig. 5. It is important that concepts of this sort are developed, and they may well give new insight into building forms, on the other hand the techniques which give rise to them are not related to the needs of actual design where analysis must lead, not to an ideal, impracticable form but to the best solution within the constraints imposed by other aspects.

In the environmental field, although the optimum form for each aspect of design may be established it is difficult to produce, by analytical means, any overall optimum since the weighting between the different aspects cannot be established. This means that it is difficult for workers trained in research disciplines to make significant progress by conventional research techniques. Because of the rapidly developing technology of building and the changing requirements it is impossible to depend for information on study of actual buildings. Results will be available too late and in many cases will not be relevant to the current problems. The building design professions have a responsibility to meet the problems of form, structural materials and function in the context of overall design and to define and sponsor studies which will contribute to good overall solutions.

A broad pattern of efficiency and economy in the selection of building forms for environmental control is easy to achieve. The precise evaluation of the cost of one solution against another in the context of design is not feasible with present techniques of analysis. There is, however, an aspect of design which gives clearly demonstrable economic returns in all environmental fields. It is designing efficiently to reduce the size of the building. Diagram 6 shows how little the variation of form affects heat loss in comparison with the major effects of increasing the volume. Costs of lighting, ventilation and sound insulation are all reduced by reducing the size of the building and this reduction is generally more than directly proportional to the reduction in size. It is not fruitful to embark on a discussion of whether buildings designed today are ideally

Fig. 6

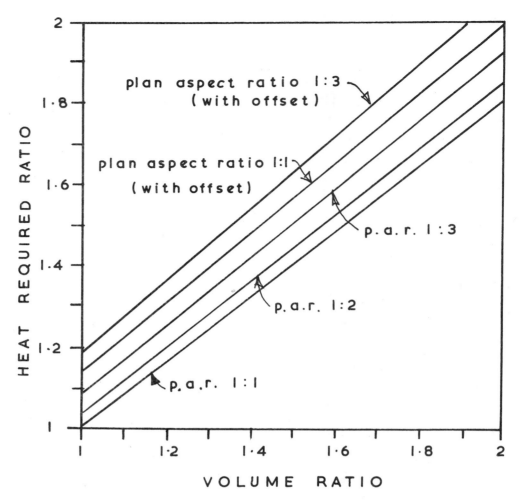

Chart showing the variation of heat loss with change in volume and change in shape. Note that for a given volume major variations in shape have only a minor effect on heat loss whereas heat loss increases approximately directly with the volume. (Original published in the Architects Journal). Curves show how the heat required varies with volume for various plan forms. (Takes into account 20 GJ per season heat from occupants, cooking, water heating, etc. Solar gain has not been taken into account).

economical of space. Spatial provision is not a fixed functional need. It is always a balance between what would be desirable and the cost of providing it, at present the judgement is based on the initial cost of the building.

Costs in use are employed to decide between different types of installation and finishes but not normally to influence the space provided. If this type of study is carried out it is likely to demonstrate that the capital savings arising from reduced size are modest in comparison with the savings from reduced environmental control over the life of the building. The potential savings will depend upon particular cases. The method however, enables a rational approach to be made to economy of environmental provision. Form, structural materials can be selected, appropriate to the needs of the design, on the basis of judgement, aided by the indication from research studies. Minimum volume forms a common basis for assessment of alternatives and progressive design can be undertaken, with the object of reducing the size of the building to a minimum.

DISCUSSION

M. Curtis (University of Reading)

I would like to agree with Professor Burberry and Mr Mitchell on the importance of getting the building right before starting to think about the services. I would be interested in comments on the BRE work by Milbank and others* on design risk applied to non air conditioned offices in summer time. It would appear to me that a similar technique could be applied to heating and air conditioning systems and that by accepting some risk of failure, reductions in the size of plant could be made and hence, presumably, the capital and running costs would be reduced.

Professor P. Burberry

I very much commend this type of approach. The building design professions are still living in the 19th century deterministic world where cause leads inevitably to effect, and finite safe values could be determined. In real life this is by no means the case. The majority of the phenomena, that one meets in nature, are probabilistic in their behaviour. There is always the chance of a rain storm of greater intensity or of higher or lower temperatures than those recorded. The situation is being recognised slowly. One interesting item in this development is the application of statistical data to standards of wind loading of buildings.

You do not have to go back very far to reach a time when everyone thought that there was a maximum wind speed and if this was used as a design standard, the building would be safe and the whole problem totally settled. Of course, that is not the case at all, there is always some possibility of higher wind speeds and the design standard chosen is not one that is totally safe but one which has reduced the risk to an acceptable level. There are more sophisticated problems related to meteorological factors which give rise to critical conditions when they occur in combination. Very little work has been done on this problem and many buildings are designed for two sets of the worse conditions which in practice, are unlikely to coincide. It is important that designers appreciate the probabilistic nature of the environment around us and make judgements on design standards in this way, rather than aiming for 'safe' standards which would be both extremely expensive and illusory.

H. G. Mitchell (H. G. Mitchell & Partners)

The judgement we must exercise is to determine the confidence limits of what we are doing, using different interest rates, pay-back periods, etc in the analysis the varying importance of different aspects can then be seen. The problem when applying this to building design is the huge one of actually predicting what you are doing. As Professor Burberry showed by graphs of calculated heat gain (Fig 6), you get all sorts of answers. We have to devise methods of making the predictions we need in the short time usually available.

R. Cullen (Architects Design Group, Derby)

I have a question for Professor Burberry which seems to be at the nub of our dilemma. When I was his student some twenty years ago he taught me to admire Le Corbusier and to believe in the premise that form follows function. It seems to me that we might be saying now that function follows form; the form being determined by cost, regulations etc. We used to have preconceived ideas about what buildings should be like as did Le Corbusier and that led us into all kinds of traps, but at some stage or other, we have to determine form and it seems to me that over those 20 years Professor Burberry has gone from the unite d'habitation to the thatched cottage. I would like to ask how does he now determine form if it is not a preconception.

Professor P. Burberry

I do not think my point of view on this has changed. If function is ignored in determining the form of the building, the result will be sculpture not architecture. It is implicit in the requirements of any building that it serves a purpose and clearly this must have a major part in determining the form. There is often a wide range of functional possibilities and it is neither inappropriate nor difficult to allow aesthetic considerations to influence decisions. The skill of the architect is to be able to achieve a form which satisfied functional requirements, satisfies aesthetic needs and is economical. It is, however, important to realise that the whole building problem is becoming very much more complex than it was when Corbusier was writing. There are many more factors to be taken into account than existed fifty years ago and this clearly includes environmental control. Many of the ideas for buildings dating from the 1920's would have been uninhabitable because environmental considerations had not been appreciated.

K. A. Wahab (University of Ife Ife, Nigeria and University of Reading)

I agree with Professor Burberry in his highlighting of the importance of orientation in the achievement of comfortable conditions in the interior of buildings in the tropics. Large eaves have been successfully used in many designs not only to reduce the direct radiation on walls, but also to protect some wall forms which may be affected by driving rain and excessive winds. The mud walls commonly found in the tropics certainly require such protection, even when they have been rendered by sand cement mixtures.

Apart from these two methods of obtaining a comfortable internal environment in the tropics, two other devices have been identified by Givoni.* The first involves the selective positioning of window openings on the exterior walls. Instead of providing one large window width on the wall, the same width, when divided into two or more parts, induces far greater air movement and does not rely on orientation for maximum performance. Observations so far made in some buildings in Nigeria confirm that air movements are greater in rooms of this type. So often some designers in the tropics recommend air conditioning installation for buildings which cannot be advantageously positioned to trap south westerly winds or cross ventilation. Givoni also asserts that by carefully adjusting the width of the walls between the windows it is possible to overcome these disadvantages.

The second method of inducing air flows in the room is achieved by erecting projecting vertical walls at the edges of the window openings. I wonder if Professor Burberry can enlighten us further regarding his experience in the performance of these devices.

On page 120 of the proceedings reference was made to what the author has called 'exploiting the possibilities of the element at the designer's disposal'. This is particularly relevant to developing countries and the type of aid they should receive.† So often designers from the highly developed countries come with preconceived ideas

Footnote

*Milbank N. O. *et al.* 'A new approach to predicting the thermal environment in buildings at the early design stage.' BRE Current Paper CP2/74.

of what is right for these countries without examining possible locally tried and successful solutions. There are as yet many more resources to be exploited in the developing countries and although these resources at present may not quickly contribute to cost reduction, in the long run, they may well do so. To go for a lofty design for its own sake or a building with a lot of glass, are methods which are really not efficient. It would seem sensible for designers wishing to practice in the developing countries to seek to know the rationale behind some designs, modifying them when such modifications can be justified, but not rushing to demolish these ideas and concepts merely to push forward a new one. Buildings usually last much longer than their designers even when they fail abysmally.

*Givoni, B. Proceedings International Conference on *Housing for the Developing Nations*. Tel Aviv, Israel, December 10–17, 1974.
†Wahab, K. A. (1976) *'Effective international assistance on housing for the Developing Countries.'* Ekistics, Vol 41, Number 242, January 1976, pp 27 and 28.

Professor P. Burberry

The effectiveness of vertical louvres for sun control on buildings, is determined by the geometry of the building, its orientation and latitude. The best place at which to control sunshine is before it penetrates the building. External louvres can be very effective. This is demonstrated in our own latitude by the effectiveness of horizontal projections over south facing windows which cut off the sun in summer and allow it to penetrate in winter. In other circumstances projections of that sort would not be particularly appropriate and with sunshine coming from an angle across the face of the building vertical louvres can be very useful.

The point about designers going to other countries and producing buildings not knowing about the local conditions is, unfortunately, very true. Many tropical buildings have been designed as if they were in Northern Europe. We do not have to go to the tropics to see the problem. Not so long ago buildings were sited and designed by people with considerable local knowledge whereas nowadays, clients with remote head offices, commission architects, who also have no local knowledge. We should take much more account of the detailed micro-climate around buildings.

I am worried about the suggestion that the form of tropical buildings should be governed by the need to improve air movement. Buildings of that type have been built but the method was appropriate to the days before electricity was freely available. It is extremely expensive to indulge in air conditioning, but to increase air movement is very simple to achieve with a small electric fan and much more effective and cheaper than any modification of the form and fenestration of the building.

T. Smith (Steensen Varming Mulcahy & Partners)

I was interested to see the relationship given by Professor Burberry of the calculated present value (CPV) of fabric and services. He quoted CPV of £120/m² for fabric against £80/m² for services based on a 10% discount rate. What was the capital value of services and fabric contained in this calculation and what is the source of his information?

Professor P. Burberry

The figures of capital value of fabric and environmental services involved in the discounted cash flow analysis were £115/m² initial cost for the fabric of the building and utility services and £35/m² for environmental services. Mr Smith said that the present values for the fabric and utility services on one hand and the environmental services on the other came out at approximately the same figure. This was, of course, the point of presenting them. I am grateful to Mr Smith for underlining the significance.

K. R. Herbert (University of Sheffield)

In his paper Professor Burberry referred briefly to the dichotomy which exists in building design, between, on the one hand the difficulty of the specialist services designer to appreciate the total design problem, and on the other hand, by implication, the limitations of the architect to fully comprehend and effectively integrate the specialist advice available to him. Would he care to develop this problem a little further and say, whether in his opinion real progress is being made in closing the technical credibility gap both in education and practice?

Professor P. Burberry

I regret to say that my observation leads me to the conclusion that there is very little professional development going on. Our problems stem from the 19th Century when the building design professions were established. Most of the professions maintain, in effect, exactly the same attitudes that they had at their formation. They retain the same ethos and the same sphere of responsibility. Indeed in the mid 20th Century we could tackle building problems involving 19th Century technology extremely well. What we cannot deal with effectively are the problems of the 20th Century. This can be exemplified by reference to much of the discussion at this conference.

A major problem was the thermal capacity of buildings. At the moment, nobody is responsible for the design of the thermal capacity of buildings. Architects do not do it because it is not something which traditionally they have taken into account and heating engineers do not do it. The engineers job is to design the installation taking into account the performance of the fabric, not to design the fabric itself. It is a very important aspect of design and nobody takes it into account. The reason is quite simple. Thermal capacity was not a 19th Century problem. Buildings had massive external walls with comparatively small windows and, if it could be afforded, a low pressure gravity hot water system or, if not, open fires. No decision about thermal performance of the building was required from anyone. Precisely the same argument could be applied to a number of other important design problems.

Very little is going on to solve this. The professions are separated and the educational system is also operating in isolated groups. To my mind the solution is that we should have an Institute for Building Design and that building designers should be educated in parallel courses in educational establishments devoted to building. I am afraid, however, that this is a very long term view and while there are encouraging signs, there is little evidence of anything happening on a significant scale at present.

Chapter 10

Building Services System Selection

H. G. Mitchell

We heat and air condition buildings so that people can use them and be comfortable in them. An understanding of comfort and how it might be created is basic therefore to building and services design.

Comfort and Buildings

The factors affecting thermal comfort are reasonably well known and understood: the air temperature must be well controlled; the relative humidity may vary over a quite large range, restricted more by static electricity problems rather than anything else; the air movement must be random and fairly low; the radiant temperature should not have extremes and should be within a few degrees of the air temperature; and there should be sufficient ventilation to keep the environment 'fresh'.

Perhaps the over-riding factors, however, are air and radiant temperatures and when the achievement and maintenance of these is considered the importance of the building is immediately recognised. If the thermal efficiency of the building is poor, low radiant temperatures will obtain and the difficulty of off-setting large heat losses with high heat inputs will cause interactive convective currents which are not conducive to comfort and generally lead to complaints of cold feet and warm heads: the hardy perennials.

Thus, comfort is achieved via the building and the design of the building and the heating and air conditioning system must take this into account. The over-riding factor, however, is the thermal efficiency of the building and unless this is good the opportunity for providing comfort at a reasonable energy cost is poor—a point which after all the research and discussion of the last few years should be well understood. This is the philosophy of Integrated Design, with the proposition that the building must provide the bulk coarse control of the climate, with services providing only fine tuning.

Energy Conservation

Heating via the building means that mass will exercise a strong influence and will confine the limits for intermittent heating and energy savings. The discussion of building mass has been going on for many years and has produced much theorising on the benefits of 'low' mass buildings. But in practice no such buildings exist, and all must be rated moderate to heavy, except for some very special cases, usually domestic, using insulating internal linings—vide Dufton's experiments in the '30's.

This realisation (that all buildings, especially when the contents are included, are thermally, relatively heavy) immediately clears the way for some conclusions.

1. If comfort is to be achieved (for even short periods) the building will need heating for a relatively long period, and therefore the heating cost will be comparable whether the building is used for short or long periods.

2. Energy saving claims made for ordinary time-switched intermittent operation are unlikely to be realised (and it is highly probable that the claims made in the past are for systems which have not provided comfort, especially during the early hours of occupation). More sophisticated controls are now, however, available but whilst these may ensure comfort in cold weather and economy during mild weather, the

overall savings are unlikely to be great. Nevertheless savings, perhaps as high as 15-20%, should be possible.

3. The design of the heating system should whenever possible bypass the building, particularly for very short term occupation.

4. The major contribution to energy conservation can only come from an increase in the thermal efficiency of the building, and this should be a central design team aim.

5. After the above options have been exercised, the only remaining energy saving measure that can be utilised is the reduction of standards.

Heating Systems

In the past, some heating systems (notably panel heating) were designed in an attempt to heat the fabric of the building directly and so promote comfort, but these systems have largely fallen into disuse because of the poor thermal performance of modern buildings (with which they cannot cope) and the economic factor of high capital cost.

Thus, capital cost and the need for high output has led to the almost universal employment of steel radiator, natural convector and forced (fan) convector systems for space heating in commercial buildings, and whilst these may not be ideal, they do provide reasonable solutions and all, when allied with good hydraulic design, provide fairly responsive systems amenable to both thermostatic and intermittent control with subsequent modest energy savings.

The same philosophy is also followed in many industrial buildings, but in some, particularly those of large volume, direct heating of the individual is attempted by a variety of radiant means, and within the demands of the industrial environment, these seem reasonably successful.

Air Conditioning Systems

In air conditioning design the influence of the building is even greater than with heating, and much closer co-operation is demanded between building and service designers.

The basic principle, however, of dealing with loads as and at the place they arise is the same, except here the load has two elements, heat and moisture, and with equipment which has very distinct characteristics when dealing with these loads. Indeed, most equipment has been developed to deal with a load of a certain type, and this gives the first lead to systems selection.

The first and dominant building elements influencing air conditioning design is the window and the way in which solar heat gain penetrates it. With unobstructed fenestration, the sun's rays are not converted to a heat gain until they strike an object. If this is an occupant, direct warning (or over-heating) is immediate, but there is a considerable time lag when the receiver is the building fabric or contents. Depending on the intensity of the sun, and the size of the window, however, the air and radiant temperatures of the space may be so raised (greenhouse effect) that the resulting environment is intolerable. Air conditioning can really only deal with the air temperature side of this problem, leaving the radiant effect which still may make conditions intolerable. In addition to this, window glare may contribute an intolerable visual problem, and the usual solution is to provide some form of internal blind.

With most types of blind this immediately changes the character of the solar gain to a quick acting air temperature effect in the window plane, and induction and fan coil systems with sill outlets were specifically developed for this situation, especially when large windows create correspondingly large heat gains, in which case under-window heating and cooling capability must be installed if comfort is to be provided.

The second element influencing design is building depth, and indeed this can often dictate the need for air conditioning: deep buildings (especially those with internal rooms) must be air conditioned. But with deep buildings, and even those with the problems created by windows described above, there is a change in load type, insofar as the moisture component becomes more prominent and systems capable of dealing with this must be used if the space humidity is to be kept to a reasonable level. On the other hand the load is usually always undirectional, that is to say either more or less cooling (after perhaps an early morning warm-up period). This may lead to some design simplification, but clearly all-air systems are indicated allied with a ceiling supply lay-out, and therefore duct accommodation within the ceiling void is necessary.

Energy conservation has already been mentioned in relation to the control of the heating system in relation to the characteristics of the building. The problem resolves into obtaining an optimum between system and building. With air conditioning, this optimum is not only more difficult to obtain (because of the twin functions of heating and cooling), but air conditioning is more inefficient (in terms of energy) in obtaining a condition, and the result is an energy usage orders of magnitude higher than simple heating systems. On simple energy grounds, therefore, the application of air conditioning should always be closely questioned and thoroughly justified.

Conclusion

We may conclude that in comfort design, the building is certainly as important as the system and that sophisticated controls will not lead to great advantages over simple ones. Indeed, any system design which relies heavily on automatic controls either for its successful operation or to achieve projected substantial energy savings, should be viewed with distrust.

I would finally like to quote from a paper written two years ago[1], where I said that *"quality can only be achieved by a committed team under strong leadership, which will ensure that all problems are faced honestly and explored properly. Quality is not the outcome of spending more money than normal, nor of considering elemental costs separately. It is achieved by an intergrated team creating an integrated design."* We see no need to modify that statement today.

REFERENCE

1 Mitchell, H. G. and Leary, J. "Environmental Quality and its cost," chapter in "Integrated Environment in Building Design" Ed. Sherratt, AFC. Applied Science Publishers (1974).

DISCUSSION

L. A. Cole (Andrews-Weatherfoil Ltd., Slough)

I was surprised by Mr Mitchell's comment that there were no advantages to be gained by intermittent heating. A 25% saving was claimed in a Ministry of Education publication published long after the war, a result which was supported by tests of ours in France in the mid sixties in which we compared the performance of fan convectors and radiators and achieved very similar results.

Mr Mitchell in his paper referred to the importance of the radiant temperature making no reference to environmental temperature. I have certainly found the latter concept difficult to use but its affect is to redistribute heat input in heating and air conditioning system calculations and would presumably overcome some of the problems you referred to about control. Professor Burberry (pp 117–124) has mentioned the advantages of heavyweight structures in Africa and lightweight structures in environments with high humidity and I wonder what he feels about lightweight versus heavyweight structures in the UK climate.

H. G. Mitchell

I certainly did not mean that there is no point in having intermittent control. There is no question that intermittent control saves energy. What I did say was that I do not think much more will be gained additional to what can be obtained using a simple time switch, probably reset once or twice a year. The way intermittent control has been used to save energy is by NOT providing comfort at the time people occupy the building, so that the building is cold until 10.30 in the morning, but is that what we are trying to achieve?

On environmental temperature—I find it complex and that is why I did not mention it, it is a concept purely for use as a device for doing calculations. I cannot therefore see how the use of that device is going to change what happens in the building itself.

Professor P. Burberry (University of Manchester)

The appropriate balance of the thermal capacity of buildings is one of the important problems of the present time. The recent history of school buildings demonstrates the matter very clearly. Mr Fowler who was then Managing Director of Weatherfoil, demonstrated around 1950, that if a fast responding heating system and a fast responding building (ie lightweight) were used, a substantial fuel economy could be achieved during the heating season. I believe that the Ministry of Education Building bulletin on this subject was based on his work. The idea had dramatic and widespread influence. The type of construction and system of heating was very widely used, without taking any account of summer overheating which came upon designers as a shock. The problem became well known very soon, but there has been a 25 year time lag before the Department of Education and Science issued specific guidelines to influence designers towards the consideration of controlling summer temperatures.

This gives a good illustration of the tremendous inertia existing in the building design field. Many designers became conscious that for summer conditions, high thermal capacity had advantages and one does even come across people who have got their thermal facts somewhat mixed, advocating high thermal capacity as a solution to the energy problem. The situation is that maximum economy in winter will be given by a lightweight building with fast reacting control. To achieve maximum comfort in the summer without expensive mechanical installations, a much more massive building is required. No one knows what the best plan is. This is something which would clearly be the subject of urgent research on a sufficient scale to produce answers in quite a short period of time. In the domestic field alone, we build 300 000 houses a year without having any standard for energy conservation.

Dr D. Fitzgerald
(University of Leeds)

I speak in defence of the environmental temperature. It is an extremely valuable design device. Take the maximum and minimum temperatures prescribed by law, and those given in the IHVE Guide for comfort, and assume that the figures given in the legal limits *are* environmental temperatures. From these ranges, choose a representative temperature and use U values and other data from the IHVE Guide, then heat loss calculations based on these figures will correspond more nearly to the truth than previously. This is because environmental temperature takes account of the fact that for moderate temperature differences heat transfer by radiation is twice as important as heat transfer by convection. To simplify one might ignore convective heat transfer, and thus air temperature and consider only the radiant temperature. In the past we did the opposite, and now we are somewhere near right. The concept of environmental temperature enables us to deal with differing surface emissivities (0·9 for most surfaces, 0·1 or less for polished metals). It enables us to calculate more accurate wall surface temperatures (vital for the prevention of condensation), and its use enables us to tackle the problem of cold radiation discomfort.

Chapter 11

North American Experience of Cost Control and Building Performance

R. Flanagan

INTRODUCTION

Before attempting to comprehend the intricacies of the North American construction industry one must consider the size and structure of the country, few people in Britain are aware of its vastness for instance New York to San Fransisco is approximately the same distance as from London to Cairo. It is, therefore, misleading to make comparisons of construction prices between our two continents because of differing:

    (a)   construction systems and techniques;

    (b)   construction regulations and standards;

    (c)   cost of labour and materials;

    (d)   design requirements;

    (e)   regional variations;

    (f)   international currency exchange rates which are subject to fluctuation on a daily basis.

General Data

One cannot consider one state within America as being typical of the whole country, differing climate and construction regulations must be considered when working within one area. The size is difficult to perceive, for example the total area of the USA is 3 611 211 square miles, and into the state of Texas one could fit the British Isles, Norway, Belgium and Portugal and still have room to spare.

The population of the USA is in the region of 216 000 000 and Canada 26 000 000. The climate varies with location. The west coast experiences 26·7°C and 32·2°C (80–90°F), with mild, short winters. The east coast, particularly in the vicinity of New York and Washington, has hot, humid summers with cold winters. The southern part of the country experiences very hot and humid summers with cold winters similar to those in Britain. In Canada, in Quebec, the temperature in winter frequently falls to 0°F, but this will not stop construction work.

Unlike the British construction industry, which approximately is 55% public sector work and 45% private sector, the Americans have a much greater proportion of private development. Table 1 shows construction volume estimates for 1972-1974, highlighting the relatively low percentage of public sector work. The construction industry is not being used as an economic regulator to the same extent as its UK counterpart.

Currently the US is experiencing an inflation rate of approximately 10% per annum with unemployment at 8% of the working population, however in the construction industry the percentage is nearer 16% of the work force. Despite these figures the contractors do invest heavily in plant and have confidence in forward planning.

Table 1

Construction Volume Estimates
(1972-1974)

	Total New Construction (Millions of Dollars)		
	1972	1973	1974
Private Construction			
Residential buildings	54,186	60,084	63,132
Non-Residential buildings			
Industrial	4,676	6,108	6,853
Commercial	13,462	16,259	16,865
Religious	884	935	890
Educational	968	901	871
Hospital & institutional	3,172	3,426	3,478
Other	914	1,306	1,152
Total non-residential	24,036	28,935	30,109
Farm Construction	902	960	996
Public Utilities			
Telephone & Telegraph	3,283	3,679	3,783
Electric light and power	8,022	9,427	10,165
Gas	1,520	1,746	1,841
Other	750	866	917
Total public utilities	13,575	15,718	16,706
All other private construction			
	941	922	922
Total private construction	93,640	106,619	111,865
Public Construction			
Buildings			
Housing and redevelopment	875	955	993
Industrial	534	589	619
Educational	5,720	6,110	6,219
Hospital	1,008	1,042	1,026
Other	3,363	3,719	3,919
Total buildings	11,500	12,415	12,776
Highways and streets	10,448	10,350	11,204
Military facilities	1,080	1,375	1,250
Conservation and development	2,172	2,300	2,130
Other Public Construction			
Sewer systems	1,702	1,918	2,015
Water supply facilities	1,076	941	967
Miscellaneous	2,218	2,635	2,718
Total other public construction	4,996	5,494	5,700
Total public construction	30,196	31,934	33,060
Total new construction	123,836	138,553	144,925

COST CONTROL

Design

The structure of the architectural profession in the USA comprises a small number of large international practices located in the major cities, large by American standards being five hundred plus, and a substantial number of small practices situated throughout the country. It is common practice for the big architectural firms to undertake the structural and environmental service engineering design 'in-house'.

The constraints of professionalism do not exist in America in the same way as the UK, there are no fee scales and each firm negotiates a fee for each project; normally on larger schemes the client puts the design work out to bid. To survive in this competitive field architects must be commercially minded, the designing of buildings is a business and it is recognised as such; an increasing number of firms have a financial interest in the commercial buildings they design.

It is difficult to generalise on fee charges but it appears that fees for the same service are as much as 50% lower in the USA than in the UK, inevitably this leads to the questions why, and how is it possible to make a profit with salaries being higher than in Britain. The keywords are competition, efficiency, a willingness to mechanise and economies of scale, wherever possible computerised techniques are employed, the role the computer plays in US industry is much greater than in Britain.

Price Prediction and Cost Control

The quantity surveying profession as such does not exist in North America, however the principal duties that the quantity surveyor performs in Britain are undertaken by people with differing job titles. There are many firms in existence that specialise in the field of cost consultancy encompassing project management, construction cost management, quantity surveys, etc, Table 2 shows the services offered by a cost consultancy organisation. The business strategy is that they have expertise in the field of cost and they attempt to diversify and offer as wide a range of services as possible.

Table 2

Project Management
Construction Management
Building Construction Cost Management
Building Design Cost Management
Feasibility Studies
Value Engineering
Energy Requirement Analyses
Life-cycle Costing
Cost Estimating Services
Computerised Cost Estimating Services
Critical Path Method Scheduling
Component Depreciation Analyses
Investment Tax Credit Analyses
Preparation and Production of Proprietary Manuals
Value Appraisals for Buildings, Property and Equipment
Loss Evaluation Appraisals for Buildings, Property and Equipment
Quantity Surveys, Quantity Take-Offs, Schedules of Materials, Schedules of Quantities
Computer Software Development involving but not limited to Programmes, and Systems for Cost Estimating, Appraisals, Accounting, Billing
Computerised Budget Estimates
Cost Research
Cost Control Automation

There are no agreed fee scales for cost control services, competitive bids are usually sought for the service.

There is a growing awareness in North America of price over-run on contracts and in the past five years the cost consultants have experienced a rapid increase in demand for their services. Hitherto the economy has been so buoyant that cost has not been the paramount feature to the client. Architects were told to design buildings, they were then asked how much was the construction cost, now the client will say I have X thousand dollars, what will the money buy me.

The American Institute of Architects have responded quickly to this changing pattern and have produced a publication 'Mastercost', which is intended to provide cost planning data for architects. Cost planning is recognised as being a valuable cost control tool and on larger schemes design cost planning is undertaken. The elemental breakdown is divided into three levels. In practice Mastercost will operate in a similar fashion to the UK Building Cost Information Service, where organisations subscribe to the service and receive analyses on a regular basis. Unfortunately American contractors are reluctant to break down their lump prices into a trade or elemental form, they feel any disclosure of prices will weaken their competitive position. The only pressure which is being exerted is on certain government contracts where the client has insisted that in order for the bid to be considered for acceptance an elemental breakdown of the bid price must be provided.

COMPUTER APPLICATION IN THE COST FIELD

Budget Analysis   Considerable research has been undertaken in the field of computer application to the overall estimating process.

The larger architectural/engineering firms have estimators on their staff, their function is to provide budget prices for schemes both by computer and manually and to evaluate tender bids.

There are various levels of computer application to the estimating process. One system is described here; it should be understood that many cost control/cost engineering firms specialise in providing this service to owners, developers and designers.

At the conceptual design stage a budget analysis programme is available, for a fee of $20 using a modem link, a conversational programme will ask the following questions:
   (a) What is the building type—eg office, factory, etc?
   (b) What construction quality is required—high, medium or low?
   (c) What is the postal zip code? (This gives the computer regional location).
   (d) What is the gross area of the building in square feet?
   (e) What is the fully developed site in acres? (This is required to evaluate a price for external works).
   (f) How many stories above grade?
   (g) Is the building air-conditioned?
   (h) Are sprinklers required?
   (i) What is the predominant superstructure?
      (1) Concrete
      (2) Steel bar joist
      (3) Structural steel
      (4) Wood
      (5) Other
      (6) Not yet known.
   (j) What inflation factor is required?

The computer has on file over two thousand project cost analyses, it will give a print-out of a budget analysis by locating buildings of a similar type to the one planned, average their costs, make adjustments for regional variation in prices and add an inflation factor. The computer recognises 106 different building types and 300 US postal zip code areas.

The print-out for a proposed four storey, fully air-conditioned office building in Chicago is as follows. (Metric data has been added subsequently).

Budget Analysis—Office Building

Element Description	Total Cost $	Cost per sq ft	Cost per m²	% of Total
Foundations	36,800	1·84	19·81	4·2
Substructure	57,980	2·60	27·99	5·9
Superstructure	112,010	5·60	60·28	12·7
Exterior closure	129,260	6·46	69·53	14·7
Roofing	12,190	0·61	6·57	1·4
Partitions	64,400	3·22	34·66	7·3
Wall Finishes	10,810	0·54	5·81	1·2
Floor Finishes	25,530	1·28	13·78	2·9
Ceiling Finishes	20,010	1·00	10·76	2·3
Specialities	25,300	1·26	13·56	2·9
Conveying systems	39,330	1·97	21·20	4·5
Plumbing	33,350	1·67	17·98	3·8
Fire Protection	23,690	1·18	12·70	2·7
HVAC	109,250	5·46	58·77	12·4
Electrical	128,800	6·44	69·32	14·6
General Conditions	46,460	2·32	24·97	5·3
NET Building Cost	869,170	43·46	467·79	98·8
Equipment	10,350	0·52	5·60	1·2
GROSS Building Cost	879,520	43·98	473·39	100·0
Sitework	35,103	1·76	18·94	4·00
CONSTRUCTION COST	$914,623	$45·74	$492.33	

The computer time was used for the exercise was 6·3 seconds and the total time involved was under 2 minutes. The UK practitioner would probably feel sceptical about the validity of the output data, but used as a tool this kind of development could prove valuable and be economic to operate.

Estimating Building Costs

Estimating has previously survived as an art, the typical estimator, who is so valuable for his technical expertise, spends a large part of his time in performing mundane repetitive tasks. This has been recognised as a waste of manpower and the computer has been harnessed as an aid to estimating.

There is no Standard Method of Measurement of Building Works in America, the estimator will take-off the quantities in accordance with his organisation's stipulated procedural measurement pattern. The advantages of this is within large contracting organisations where the measurement will be compatible with the cost accounting coding system, it is therefore possible to keep close check upon estimated, actual and projected costs for specific items.

The philosophy adopted is that each operation within the construction process has a constant labour and material content, allowance is made for regional variation, inflation, etc. From standard lists there can be automatic generation of labour, material and plant requirements for a project, the estimator has the facility to add an over-ride factor for any of the items should he feel the constants are inadequate for a particular item.

Standard labour and material costs are stored on file, Table 3 shows part of the labour wage rate file. The estimator is given the current '20 city average labour rate' for the region and he must insert, using his expertise, the wage rates he feels are applicable for the scheme under consideration. The same principle is applied to the basic material price

135

file, from this information the computer calculates the adjustment factor. Table 4 shows the print-out for the foundations showing measured items and Table 5 gives the division-item summary, which is a summary of some of the measured work. Table 6 sets out labour resource requirements and Table 7 the crew day analysis.

Developments in Britain have taken place along similar lines, but these systems are commonplace in the US and are operated on a commercial basis.

Table 3

Code	Trade Name	Abbr	Local Rate$	20-City Ave$	Adj Factor
11	Asbestos Worker	AW	11·77	9·88	1·191
12	Bricklayer	BL	13·09	9·88	1·325
13	Carpenter	CP	12·07	9·49	1·272
14	Cement Mason	CM	11·78	9·36	1·259
15	Electrician	EL	12·44	10·32	1·205
16	Glazier	GL	10·82	9·07	1·193
17	Labourer	LA	9·82	7·46	1·316
18	Lather	LH	11·82	9·48	1·247
19	Oiler	OL	11·23	8·27	1·358
20	Operating Engineer—Hoisting	HE	12·83	9·82	1·307
21	Operating Engineer—Excavating	EO	12·83	9·57	1·341
22	Painter	PA	10·58	8·74	1·211
23	Pipefitter	PF	12·47	10·44	1·194
24	Plasterer	PS	10·51	9·33	1·126
25	Plumber	PL	12·47	10·41	1·198
	Price of Concrete (Cubic yard)		25·05	18·00	1·392

Table 4

FOUNDATIONS

Crew	Output per day	Labour $/Unit	Material $/Unit	Component
2 EO 6 LA 1 OL	300 Cu Yd	1·92	0·82	Wall Footing Excavation and Bkfl
3 CP 1 LA	300 Sq Ft	0·96	0·25	Wall Footing Forms
1 RI	900 Lbs	0·09	0·22	Wall Footing Reinforcing
1 LA	12 Cu Yd	4·97	21·00	Wall Footing Concrete
				** Foundation Walls—Footings
				** Foundation Walls
4 CP 2 LA	400 Sq Ft	1·06	02·26	Wall Forms
4 CP 2 LA	360 Sq Ft	1·18	0·28	Wall Forms 10′—18′ high
4 CP 2 LA	310 Sq Ft	1·36	0·29	Wall Forms 18′—24′ high
4 CP 2 LA	280 Sq Ft	1·51	0·30	Wall Forms—intricate
1 RI	805 Lbs	0·10	0·22	Wall Reinforcing
1 LA	7·50 Cu Yd	7·96	21·63	Wall Concrete
1 CM	325 Sq Ft	0·23	0·02	Rubbed Concrete Finish
3 BL 2 LA	330 Sq Ft	1·08	0·74	Block 8″
4 BL 3 LA	420 Sq Ft	1·18	0·98	Block 12″
1 BL	400 Sq Ft	0·20	0·04	Parging
1 CP	90 Ln Ft	0·84	0·34	Construction Joints
1 CP	110 Ln Ft	0·69	1·15	Waterstop
1 CP	500 Ln Ft	0·15	0·10	Key Form

Table 5

DIVISION-ITEM SUMMARY

Description	Labour	Material	Total
	$	$	$
Site Work			
Clearing of site	1,630	1,112	2,742
Site Excavation	4,831	9,466	14,297
Excavation & Backfill	32,398	14,285	46,683
Fine Grade & Gravel	6,443	4,649	11,092
Site Drainage	667	1,534	2,201
Walks	2,068	1,881	3,949
Roads & Parking	28,898	44,919	73,817
Interior Demolition	3,731	0	3,731
	80,666	77,846	158,512
Concrete			
Concrete Formwork	4,918	928	5,846
Concrete Reinforcement	4,221	6,555	10,776
Cast in Place Concrete	19,785	32,139	51,924
Concrete Finishes	14,195	3,934	18,129
Cementitious Decking	488	641	1,129

Table 6

LABOUR RESOURCES REQUIRED

Trade	Abbr	Man Days
Asbestos Worker	AW	487·2
Bricklayer	BL	3,439·6
Carpenter	CP	3,124·2
Cement Mason	CM	2,330·2
Electrician	EL	2,243·8
Glazier	GL	19·6
Labourer	LA	5,786·1
Lather	LH	0·0
Oiler	OL	507·0
Operating Engineer—Hoisting	HE	449·9
Operating Engineer—Excavation	EO	123·0
Painter	PA	191·9
Pipefitter	PF	2,148·0
Plasterer	PS	0·0
Plumber	PL	5,095·4

Aids to Estimating

Work on mechanising the estimating process has been undertaken by organisations like the Portland Cement Association. It is possible to use one of the programmes developed by PCA in the comparison and evaluation of various reinforced concrete floor systems in multi-storey buildings. It is intended for use during the preliminary layout and design phase of a building project. The programme estimates quantities of concrete, reinforcing steel and forms for the floors and columns. The itemised quantities are extended by user-supplied unit costs to provide a cost estimate for the structural frame.

The advantage of this type of system is its speed of operation and the flexibility of immediately being able to see the effect on cost of changing the design live load, span, etc. It also provides a valuable cost optimisation tool at the design stage. The steel corporations have also structured programmes that will perform a similar function for steel framed buildings.

Table 7

CREW DAY ANALYSIS

Work Activity	Quantity	Unit	Standard Crew	Crew Days
Gravel Encasement	236	Cu Yd	2 LA	4·1
Porous Wall Concrete Pipe 4″	112	Ln Ft	1 LA	1·7
Underpinning incl Concrete	2	Cu Yd	2 LA	2·0
Permanent Forms to Back U'pinng	14	Sq Ft	1 CP 1 LA	0·2
Wall Footing Form Work	67	Sq Ft	3 CP 1 LA	0·2
Wall Forms 10′—18′ high	385	Sq Ft	4 CP 2 LA	1·1
Wall Forms 18′—24′ high	522	Sq Ft	4 CP 2 LA	1·7
Wall Reinforcing	1	Lbs		
Wall Reinforcement	967	Lbs	1 CP 1 LA	0·7
Wall Footing Concrete	1	Cu Yd	1 LA	0·1
Wall Concrete	9	Cu Yd	1 LA	1·2
Rubbed Concrete Finish	88	Sq Ft	1 CM	0·3
$1\frac{1}{4}″ \times 1\frac{1}{4}″ \times \frac{1}{4}″$ 'L'	35	Lbs	1 CP 1 LA	0·1
$1″ \times \frac{1}{4}″$ Galv Steel Grating	9	Sq Ft	1 CP 1 LA	0·1

Construction Price Indices

It is worthwhile to examine the factors that can influence American price indices. The high cost of transportation of materials is very significant and must be considered at the design stage. Table 8 shows an extract from an '80 city' price index, for adequate interpretation it requires a thorough understanding of the economic factors affecting the index, for example the figures for Anchorage, Alaska will be inflated because of the importation of labour and materials. The index for structural steel in Baltimore reflects its close proximity to the steel producing plants.

Table 8

PRICE INDEX

	New York City, N.Y.	Albany, N.Y.	Albuquerque, N.M.	Anchorage, Alaska	Atlanta, Georgia	Baltimore, MD	Chicago, Ill
Composite Index	100	96·3	89·9	104·2	94·5	101·3	95·6
Concrete Work	100	97·6	96·1	102·2	93·2	99·6	98·0
Masonry	100	94·2	90·3	103·5	85·8	99·4	93·4
Structural Steel	100	100·1	93·7	114·9	96·4	88·7	96·1
Roofing	100	103·9	88·8	119·8	111·2	104·2	99·7
Plaster	100	95·2	87·7	104·0	91·2	98·7	95·7
Gypsum Dry Wall	100	92·4	86·3	104·7	93·7	110·7	94·3
Acoustic Treatment	100	91·3	88·9	100·5	92·5	103·5	91·2
Resilient Flooring	100	91·7	82·6	98·7	94·2	101·6	91·8
Sprayed Fire Protection	100	99·2	94·2	100·4	93·6	99·0	97·5
Painting and Finishes	100	92·6	82·9	96·8	90·7	103·1	93·4
Plumbing	100	96·5	91·6	110·0	96·5	98·2	101·2
HVAC	100	97·1	86·2	109·3	95·8	99·1	96·1
Electrical	100	97·9	93·6	113·9	93·1	100·1	98·8

Labour costs vary with location, sites are classified as either union closed shop or non-union, the union schemes are usually located in the cities and all employees must hold a union membership card. When vacancies occur contractors contact the union headquarters who draft the man required. Many contractors will operate both union and non-union sites, labour wage rates tend to be slightly higher on union projects. The unions are very conscious of the need to keep productivity high with correspondingly high wage rates.

An attempt is being made to educate clients to the significance design decisions have on the operating and maintenance costs of buildings. Spiralling energy prices have brought about an awareness of energy consumption which hitherto the American client was not prepared to consider.

The problem has been tackled in a similar fashion to the British cost-in-use techniques, except on commercial schemes where the US owner requires investment tax credit advice and component depreciation analysis. Investment tax credit incentive schemes are in operation to stimulate investment in the manufacturing industries. Certain types of building will qualify for a tax credit but they must fall within certain categories, the cost consultant can advise at the design stage on the implications of the design on the tax credit allowances.

CONSTRUCTION MANAGEMENT AND CONSTRUCTION EFFICIENCY

It is not the intention to discuss here the developing role of the construction manager and project manager in the American construction industry. The construction manager or project manager acts as the client's agent with responsibility for seeing that the building is realised on time, within the budget and with its functional and aesthetic goals intact.

Change orders during the progress of construction work are to be avoided because of the cost, the client will pay heavily for the variation.

In broad terms it is possible to construct a building in America for a lower cost and quicker than the same structure in Britain. This is a very wide statement and without elaborating in depth one can only talk in generalities. The prime reasons are:
  (a) Economies of scale.
  (b) High investment by contractors in plant and mechanised tools.
  (c) High labour productivity, workers work hard because of job security and high wages.
  (d) Industrialised building techniques, wet trades are eliminated wherever possible.
  (e) Fierce competition.
  (f) Abundance of certain types of raw materials.
  (g) The contractors' supervisory staff to worker ratio is kept as high as possible, up to 1:18 in many instances.
  (h) The size of the construction market means many of the components are priced on high volume sales with low profit margins.
  (i) The marketing strategy of contractors is geared to satisfied clients and repeat business.
  (j) The climate has dictated construction systems that have speed as the criterion.
  (k) Emphasis on pre-planning.

BUILDING PERFORMANCE

The extremes of climate dictate the environmental performance required from the building, some form of air-conditioning is desirable and attention is paid to the thermal insulation requirements. In the north and east of the country double glazing is normally used. There tends to be a great deal of standardisation of component sizes which, at times, can be somewhat restrictive on design.

Attention is paid to the elevational treatment of buildings, this is a critical factor of design. The Americans are becoming increasingly aware of their built environment, for example in certain areas of California it is official policy that developers devote at least 1% of the building cost to exterior works of art.

Emphasis is placed upon safety, particularly against the risk of fire, ultimately it is the installer not the designer who must accept responsibility for the safety of the building with regard to construction and fire protection.

CONCLUSIONS

In Britain things tend to change by evolution rather than revolution, particularly in the professions.

We currently live in a rapidly changing world, the Americans appear to react and respond very quickly to change and above all they seem prepared to try new ideas.

We have very little to offer the Americans in the field of cost control, their knowledge in certain areas is far in excess of ours. We should not be too proud or embarrassed to copy some of their techniques and develop them for the UK market.

DISCUSSION

T. Jarman (University of Natal, Durban, South Africa)

A high proportion of Canadian Public Building works are financed either directly or indirectly by the Federal Government with the result that the 1969–70 cut back of government spending was devastating to the building industry and allied professions. I would like to point out that quantity surveyors are established in Canada where I worked in a multi professional team which included a quantity surveyor.

In the USA architects and engineers are, I believe, professionally orientated. They are registered by the State similar to the registration of architects in the UK by ARCUK (Architects Registration Council of the United Kingdom).

My experience leads me to question the American/Canadian attitude towards brokerage type contractors who work on a complete set of documents prepared by the professional teams. The shopping around and price fixing inherent in this system leads to a high incidence of bankruptcy among sub contractors. With the absence of quantity surveyors the procedure is quite different from that in the UK. However the time to prepare contract documents does require programming so that they can be prepared prior to bid time and many sub contractors have indicated to me that they would appreciate the possibility of tendering on a bill of quantities but it is doubtful if the brokers who assemble the price for a building would accept that system.

R. Flanagan

The first point on the question of the public sector. I would not dispute the high figures for public sector building in Canada. My figures were for the USA which is not strictly comparable with Canada. I did not say there were no quantity surveyors in Canada after all I work with an organisation with its head office there. The important point is their presence in such small numbers with no more than an estimated 25–30 firms actually practising as cost consultants/quantity surveyors.

Architects and engineers must be registered by the State to practice but in terms of qualifications it is the Universities that do the examining and are responsible for setting the standards.

A comment on price fixing and the use of bills of quantities. I can only stress what one New York contractor recently said when I asked him if he would like to tender on a bill of quantities: 'No way, in your system they all start at the same point, in our system the good guy wins'. Whether bills of quantities would be welcomed or not depends on the individual person or organisation. Many documents are prepared which are bills of a kind but the real problem is standardisation of measurement ie what is included within the measurement document.

G. Ashcroft (Balfour Kilpatrick Ltd.)

I was interested to see in the budget analysis on the office building that the electrical cost of 128 000 dollars is much higher than the heating and air conditioning cost. I have not seen this before, perhaps this is because UK electrical prices are unrealistically low.

A. J. Colledge (Department of the Environment)

I too was intrigued by the comparison of the electrician and heating trade prices because on visits to America and Canada I was not impressed by the standard of electrical work. It seemed to me that a lot of work in the USA was of poor quality, geared for a shorter life.

R. Flanagan

On the electrical installation price, the figures quoted were those from an actual live project. With regard to the quality of buildings, I feel the most significant thing that stands out is the degree of attention given to the elevational treatment on the buildings. Architects do not have the same freedom of choice, they tend to be far more restricted with the use of standard component types and sizes. There is a tendency to design the buildings for a shorter life span. The measurement of quality is a personal thing and certainly the quality of the elevational design of the buildings is generally impressive.

F. D. Manley (Telford Development Corporation)

Mr Flanagan mentioned that architects bid for their fees for the opportunity to design schemes. I wonder whether this in fact leads them to be more receptive to using standard components, since obviously this must cut down their detailing and design time. I was very concerned recently when a senior architect asked why he does not re-use good housing schemes, observed that an architect is trained to innovate and failure to innovate is backward looking and an abandonment of the essential basis of his profession. I submit that an architect who continues to innovate by the law of averages will make many mistakes and I wonder whether the American architect who is forced into the position of using standard components will cut down design time and get better buildings as a result.

R. Flanagan

Generally speaking, the quality of design in the USA is not inferior, the architectural profession strive for efficiency in every way. For example, there is a tendency to make considerable use of computers. I do not think the fee aspect restricts them nor is the quality of the design non-innovatory, competition is healthy. The American architects I have spoken to have very mixed feelings on the question of fixed fee scales. The general feeling is that if you are good, efficient and competitive you will get work. The product can often be judged on the amount of repeat business.

A. Donaldson (Wylie Shanks & Partners)

I admire very much many facets of the North American construction industry, in particular the emphasis that is placed on pre-planning. However, I really have severe reservations about selection of the design team by bidding. To mis-quote Mr Doig in his introduction 'How would you feel if you were a space pilot, not only sitting in an aircraft made of components that were supplied by the lowest bidders, but also designed by the lowest bidders". I would like Mr Flanagan to describe in greater detail the selection process of a design team.

R. Flanagan

I have only limited experience of design team selection. The client sends out a brief of the building to a number of design teams or designers chosen perhaps by recommendation. An interview takes place, as generally happens in the UK, but the over-riding features tend to be the fee aspect and the track record. The client will be very interested in previous work done by the designer, his client list and his intended approach to the scheme. A lot of money is spent on public relations and marketing. They say the product they are marketing is themselves and they market it to the best of their ability.

I speak in generalities from the point of view of quality and what the client gets at the end of the day. Design is an extremely competitive field and the importance of marketing cannot be over emphasised, some of the larger architectural practices have business development units and the marketing strategy they adopt is quite interesting.

T. Smith (Steensen Varming Mulcahy & Partners)

I am slightly familiar with a number of the American buildings Mr Flanagan used to illustrate his presentation, in particular I would like to draw attention to the Hyatt Hotel. He mentioned that this building is a commercially successful unit. From my knowledge of it, it is also a building which incorporated components used in a grossly extravagant way. Building components utilise natural resources in their manufacture, transporation and construction. It must, therefore, be recognised that in any attempt to apply the science of Terotechnology to building, we must recognise a factor which has not previously been considered. I refer to the success of the building in purely commercial terms which is totally unrelated to the efficient use of natural resources in terms of Terotechnology as we know it.

Chapter 12

Cost and Quality Data Banks in Operation

J. D. M. Robertson

Introduction

No decision can be confidently taken and no advice authoritatively offered if factual information is not available or is not used. Nevertheless, no matter how much measurement or description is available, correct advice or a correct decision will not be forthcoming unless the information is interpreted by knowledge and experience. Professional judgement and expertise are crucial ingredients of decision taking: a good information base is also an important ingredient so that the industry can keep abreast of rapid change in economic, social and technological development.

The general view is that there is already too much information—too many Codes of Practice, Building Regulations, manufacturers' catalogues, research reports. Certainly there are too many facts and too much information for an individual to remember at all times. What is now important is to know how to gain access to relevant data. Not surprisingly therefore, much effort is placed on standardisation, classification, bibliographies, and library based services.

Inevitably through the information services that I operate—the Building Cost Information Service on behalf of the Royal Institution of Chartered Surveyors and the Building Maintenance Cost Information Service sponsored jointly by the RICS and the Department of the Environment—I am involved in the standardisation of documents, classification and coding systems and the preparation of bibliographies. Our library on building economics has been built up over the past 14 years and is the basis of a photocopying service which is widely used. Books, articles and papers are continuously being added to it. However it is not these activities which are described here.

Here the concern is with data which are designed to assist the specific activities and techniques associated with design/construction economics and buildings-in-use. It deals with information and information systems which can be built around the techniques used in decision-taking, or in initiating action, or in control and accounting procedures.

Information systems

Hindsight often attributes the cause of a wrong decision to a lack of information. What should be encouraged is the establishment of information systems which allow relevant facts to flow within an organisation in such a way that decisions can be made in the full knowledge of up-to-date data and current experience.

The introduction of an information system should always be preceded by a study to find out what procedures and techniques are already in use and what information flows are most natural. The analytical study should aim to anticipate what information will be required and how much detail, how it should be presented and how frequently it will be consulted. It should also identify how the system can generate its own information, how data should be collated and how management could use the data for solving problems or monitoring performance.

Communication is a two-way thing. It involves a transmitter and a receiver and these will not work unless there is a sound channel between them and a clear signal. Successful communication achieves the desired effect; a negative communication produces nothing at all; miss-communication can lead to the very opposite happening. The success of an information system must be judged by the quality of the decisions that are made and the actions which are initiated.

Whilst management may be occupied with establishing its new objectives it also has to deal with situations which are failing to achieve existing plans. Information systems therefore which can help establish norms are essential parts of dynamic management.

Standardisation plays an important part in any successful communication system. Standardisation of instruction and definition is required for consistent analysis. Standardisation of classification and coding is required to facilitate storage and retrieval whether it be in a library or on a computer. Standardisation of presentation of data sheets is required to promote familiarity and confidence in the selected matter. In addition, a standard of discipline is required to make sure that information is automatically generated by the system and that it is subsequently used in decision-taking.

The preparation of data is so often considered to be an inessential and non-productive operation that it is not readily adopted as an automatic routine. Too much information requested out of an unspecified or general interest or for research purposes detracts enormously from a person's willingness to update records. Likewise, too much information which has to be sifted through before the relevant part can be identified is also something to be discouraged. Information must have a practical application.

There is always the question of how much trust can be placed in information. In this respect data used in building can be classified under four main headings:

(i) *Specific to a project*—
   for example—the client's brief; terms of reference; conditions of contract; production drawings; structural calculations; information only available to those engaged on the task.
(ii) *Specific to an organisation or firm*—
   for example—cost and price records on projects dealt with by the office; standard office specifications and drawings; information only available to members within particular firms and derived from a total office experience.
(iii) *Specific but supplied from outside sources*—
   for example—statements of inter-firm comparison; exchange services, e.g. cost analyses; specific information related to particular projects made available by an organisation to other organisations.
(iv) *General*—
   for example—manufacturers' catalogues; codes of practice; building regulations; information which could be applicable to any project and available to everyone.

An individual knows the background of information derived from sources within his control—in fact this constitutes his experience—and quite naturally this is the information which he most trusts. Information specific to the organisation in which an individual works will also be used readily because its validity can be ascertained from colleagues.

Standard documents such as Codes of Practice and Building Regulations are relied upon because they have been issued only after detailed deliberations by experts in the field. On the other hand, data sheets projecting the efficiency of one or another sealant or detergent which could be said to be of a marketing character, have to be carefully sifted.

Information derived from outside sources including other projects (much of price and cost information is in this category) also has to be interpreted with professional care. Information obtained from a particular project is likely to be factual but it does not necessarily show whether or not it refers to the optimum solution to the problem. Even information generated within the organisation should be checked against norms

144

calculated from as large a sample of comparable information as possible. A surveyor is more confident of his own calculations or opinions if they can be corroborated easily from comparable information, preferably prepared by his professional colleagues using a standard approach to the problem.

The information system approach has been adopted successfully by the two reciprocal information services—the Building Cost Information Service and the Building Maintenance Cost Information Service—which are unique in the generation of knowledge for total cost appraisal.

AN INFORMATION SYSTEM FOR BUDGETARY CONTROL OF PROPERTY OCCUPANCY EXPENDITURE

Property occupancy embraces such expenditure headings as redecoration, fabric maintenance, servicing, cleaning, gas, electricity, oil and other utilities, administrative costs such as laundry, porterage, security and property management, and also property insurances and rates.

Building maintenance costs alone in the United Kingdom have been estimated to be in the region of £3 000 000 per annum, not to mention the amount spent on cleaning, energy and administration. However, it is still considered to be an area of low priority by management, designers and contractors. It is a field lacking in terminology and definition, where management procedures vary enormously, and where even responsibility is diffuse, shared by company secretaries, engineers, building supervisors, and professional advisors. The survey which was conducted to examine which aspects could be helped by information not surprisingly identified budgeting and budgetary control as an area of potential progress.

A budget has been defined in the terminology of cost accounting as 'a financial and/or quantitive statement, prepared prior to a defined period of time, of the policy to be pursued during that period for the purpose of obtaining a given objective'. First, the property surveyor has to prepare financial reports of the maintenance work that is necessary, estimate the probable cost, and help fix the budget limits. Second, he has the responsibility for controlling the maintenance work within the agreed budget.

The ideal situation is where the maintenance manager and the financial controller actually attempt to balance pressing technical requirements with the normally limited financial resources. Such a system would allow the lessons learnt during the year to be used to improve the budgetary procedures thereafter. It would encourage property surveyors to improve on their technical management and for upper management to formulate a deliberate policy towards the maintenance of their buildings.

It was within this area that the Building Maintenance Cost Information Service (BMCIS) saw an opportunity to make a contribution by producing a document which could be used as the basis for more systematic budgeting and budgetary control. The document recommends principles, instructions, definitions and elements for the analysis of property occupancy cost. A standard format has been produced by BMCIS primarily to suit the system and secondly as a means of collecting and disseminating maintenance cost information. It is intended that this document, its elements and its format should be used by those responsible for property occupancy expenditure as a sensible means of:

(a) reporting to upper management at the budget estimate stage;
(b) conducting regular checks within the budget limits throughout the financial year;
(c) analysing annual expenditure;
(d) providing a guide to future expenditure.
The main principles of this standardised report are that it should:
(a) contain information which helps organisations understand their own property occupancy problems and find solutions;
(b) facilitate communications between upper management and maintenance management centred around budgetary procedures;

(c) facilitate comparison of property occupancy expenditure from year to year within an organisation and also allow comparison with the annual costs of other organisations;

(d) permit feedback of cost data which can be used for an appraisal of capital costs and running costs, that is a total cost appraisal.

The instructions which accompany the document identify the criteria which are important in assessing the significance of the analysis. They identify: building type; owner/occupier; region; date of erection; upper management criteria; budget procedures; maintenance management and operation; building function; form of construction and building parameters. The standard format includes elements or cost codes against which expenditure is expressed in £'s per 100 metres square of floor area, as well as in total.

The instructions describe what items of work should be allocated under each cost code but although the intention is to standardise practice the list can be expanded or reduced to suit individual requirements. It has been established in its present detail because it is compatible with most existing accounting systems. In selecting codes for the analysis of running costs it has not been possible to introduce a list which is totally compatible with the elements of capital cost analysis. Although this is desirable and should eventually materialise, to facilitate cost-in-use appraisals, it is not thought feasible at this stage to depart too far from the maintenance manager's concept of cost control. It is not sensible to try to introduce guidelines for a new information system if the proposals do not acknowledge the existence and importance of local conventions, many of which ought to be retained at least in the first instance. However, when maintenance costs are systematically recorded on the standard format little trouble will be experienced in producing the data needed for total cost assessments.

Owing to the cyclical nature of maintenance, expenditure has to be considered over a number of years and BMCIS has had to make provision for this in a composite analysis which compares expenditure on one building over a number of years.

Since there did not already exist a standard accounting system for property occupancy expenditure, the introduction of the system has not been easy. Many organisations have welcomed guidance on the subject but have still some way to go before they can introduce the analyses as standard practice. Nevertheless, cost information is now being collected and is being exchanged. In time, the data will be adequate to establish norms which can be used for setting budgets for building maintenance. The system will provide a means of comparing the budget with actual expenditure. It will be able to identify items of exceptional expenditure which will prompt closer examination by management. This could lead to clients to alter their instructions to their architects and to revise their policy for setting maintenance budgets and standards.

Information for capital expenditure control

The Standard Form of Contract, the Standard Method of Measurement and consistent tendering procedures have contributed greatly to establishing uniformity in communications and documentation. Bills of Quantities are the medium through which a client and a builder make their legal and financial agreements. It is also a prime source of price information. As well as being the basis of agreeing the amount of the final account to be paid by the client to the builder, the Bill of Quantities suitably analysed produces invaluable information for economic and design studies which can be used in the planning of a new projects.

For the past 20 years there has been developing the technique known as 'cost planning' which is defined as 'the systematic application of cost criteria to the design process, so as to maintain in the first place sensible and economic relationships between cost, quality, utility and appearance, and in the second place such overall control of proposed expenditure as the circumstances dictate'.

The objectives of design economy are clearly stated, and unlike property maintenance the techniques are well defined and a source of design/price information is readily available. The problem here, however, is that cost planning techniques require information to be presented in a particular way which requires considerable re-sorting of the prices contained in Bills of Quantities. Nor is the same detail of price information required at the various stages of cost planning.

To overcome this, the Building Cost Information Service (BCIS) published, in 1969, the Standard Form of Cost Analysis which sets out principles, definitions, elements and instructions for analysing Bills of Quantities to produce price information which can be used directly in cost planning techniques. The elemental lists are recommended for the presentation of cost plans and for monitoring the economic relationships of any change in design. The Standard Form of Cost Analysis was welcomed by the profession and by Government Departments and has been widely adopted for collecting cost information. It is used as part of the normal procedures in quantity surveying offices; it is the basis of pro-formas used by various Ministries; and it is the method of classification used in the cost data bank of BCIS.

In support of the SFCA, BCIS has produced three forms of analysis in different degrees of detail—the Concise Cost Analysis, the Detailed Cost Analysis and the Amplified Cost Analysis. These have been designed to match the information requirements at various stages in cost planning.

At the earliest design stage the client requires to know the order of his likely expenditure. Where design information is imprecise cost advice has to be based on as large a sample of cost information as possible. For this purpose the BCIS Concise Cost Analysis has been developed to build up ranges of costs from a large number of building contracts. The information contained in Concise Cost Analyses is relatively brief, but in bulk they provide a comprehensive guide to the costs of a wide variety of building types.

The Concise Cost Analyses identify location, function, construction form, and give a brief specification of the building as well as a brief statement of important contract information. Cost are broken down according to six group elements, namely: substructure; superstructure; internal finishes; fittings; services; and external works. Cost is expressed in £'s per square metre of floor area. The Concise Cost Analyses are presented on an A5 format and blank forms using NCR paper/card enable users to build up their own card index system within their offices. Analyses in this detail are used in the preparation of feasibility studies at the earliest conception of a building project and are used to confirm preliminary budget estimates.

A greater depth of information is required in the later stages of cost planning. The second level of analysis provides information suitable for more fully developed estimates and for establishing an initial cost plan. This analysis contains desciptive information of the project related to market conditions, contract particulars, client's functional requirements, specification, and parameters of design/shape. The group element costs of the Concise analysis are further broken down into 28 building elements. For each of these elements a record is shown of: its total cost; its cost in £'s per metre of gross floor area; its particular parameter (such as area of external walls or number of sanitary fittings); and its cost related to its particular parameter. This degree of detail provides suitable information for assessing the probable capital costs of new projects and establishing a balanced budget within which the design team can operate.

Cost planning is a technique used to obtain value for money and as such requires economic assessments to be made of alternatives—alternative design solutions, forms of construction, components or materials. A further degree of analysis is required to provide the information necessary for such cost studies and cost checks. The BCIS Amplified Cost Analysis has been developed for this purpose. The 28 building elements are considered individually in much more detail since various forms of construction can exist in one element. Information is recorded which describes the elemental design parameters of various types of construction and specification of the materials used. Costs are broken down according to the individual types of construction within the element.

Quantity surveyors prefer to use cost information derived from projects that they have personally been associated with or which have been dealt with elsewhere in their firms. This is information which they can trust for they understand it fully. They will use published information in BCIS or the technical press as a guide to their own calculations or to supplement their own information if it is inadequate in any respect. This is achieved much more easily now that all cost information is analysed, prepared and presented on exactly the same basis.

Information on economic trends

Some information is more reliable if it is processed centrally. Statistical information requires a large sample of data before it can be considered satisfactory. Most surveying firms or departments cannot call upon the necessary amount of data and they therefore have to rely on material produced by centralised statistical units. There is in Great Britain no shortage of statistical material and the relevant data are collected by both BCIS and BMCIS in their respective fields. However, there is one specific study conducted by BCIS as part of its reciprocal arrangements with its subscribers which is worth special mention. This is in connection with the calculation of a tender price index.

One of the most important statistics in the building industry is a price index. It is used in economic planning by Governments, for estimating the probable cost of new developments, for insurance purposes, for fire loss adjustments and even for adjusting rents. The BCIS Tender Price Index is not designed to satisfy all these purposes, but it is still used for all of them nevertheless.

A sophisticated statistical study was undertaken by the Department of the Environment to produce a system of analysing Bills of Quantities which compares the rates of individual items of work against a series of base rates. BCIS has adopted this method and with the co-operation of its subscribers offers a specialist service.

About 300 Bills of Quantities are collected each year from quantity surveying firms. Twenty subscribers, selected for statistical reasons by random sample, are invited each week to submit a priced Bill of Quantities for analysis. This normally leads to receipt of about six Bills of Quantities since not all firms will have a Bill which satisfies the BCIS criteria at that time. The Bill of Quantities is then processed by BCIS staff who prepare a report which is sent to the subscriber for his own purposes and is not made available to anyone else. The individual quantity surveyor is presented with a statistical report on the pricing level of his project which is of enormous benefit to him. BCIS is in a position to aggregate the individual tender price index figures and publish a quarterly series which traces changes, through time, of tender prices in the industry.

In support of the Tender Price Index, BCIS also calculates a Factor Cost Index adopting a method which follows the procedures used by builders in preparing their estimates. Hypothetical models of the resources used in various building types or in various types of construction have been produced. As changes occur in the cost of labour, materials and overheads, their values are fed into the models to produce index series of building costs (not to be confused with tender prices). The weightings of the Factor Cost Indices are at present being transferred to a computer and the system will be able to monitor the effects of any cost change, eg in steel prices, as soon as it is announced. It is important that cost and price information is communicated to practitioners as quickly as possible.

The office prepares the 'Building' Cost Information File which includes a Housing Cost Index, Measured Rates and Current Prices of Material and Labour. Surveys are also undertaken to find out the economic forces affecting market conditions in the building industry on a regional basis as well as studying the economic relationships in particular, usually large, projects and their ramifications on a locality.

Design/performance information

Inter-professional and inter-disciplinary communications need a lot of encouragement. There is much criticism of designers who pay scant attention to the problems of maintenance and properties-in-use. Likewise there is as much criticism of property owners who fail to notify architects of their experience of occupying buildings.

On the financial side, the techniques of cost-in-use and total cost appraisal have been well known for many years but seldom have they had a practical application. Many people have talked about the relationship of capital costs and running costs. Few people have made much progress in persuading clients, Central Government for instance, to allocate additional capital funds in order to reduce subsequent running costs or maintenance. However, it may be that the changing economic circumstances brought about by the rapidly increasing costs of energy will mean that the overall concept of total cost and properties-in-use will receive more attention. Certainly, in design more careful consideration will have to be given to the specification of mechanical and electrical services and their inter-relationship with the specification of structural elements such as cladding.

The preliminary BMCIS survey showed that there was little information available on the cause, the effect, or the cost of a building failure. There was little feedback of information on the ramifications of design to highlight critical design decisions. Building occupiers failed to monitor evidence on the performance of materials. Nor was there much information available on maintenance standards, deterioration rates, or life expectancy of products. There was little performance data to help decide whether to install one or another material, component or service.

The preparation of design/performance data sheets is now another of the reciprocal sections of BMCIS. A simple standard format with instructions is used by building owners to report on the problems encountered in properties-in-use so as to monitor performance information to designers. The data sheets describe the present state of the detail, element or component; forecast the likely long term effect; analyse the cause of the failure; and suggest any design or construction correction which might have avoided the situation. Suggestions are also made on what remedial action is necessary. Diagnoses of cause and effect requires a very high degree of knowledge of the reactions of physical chemistry and all data sheets submitted to BMCIS have to be scrutinised very thoroughly by building scientists before they are published.

The exchange of design/performance information is a procedure which ought to operate as a matter of course in most client/designer relationships so as to avoid the repetition of failures. This is another case where an organisation's good practice can produce valuable information which can be disseminated anonymously to other organisations. It is proving a most popular service and is a good example of the feedback of information on an interdisciplinary basis.

Although feedback of information in many fields is unusual it must not be forgotten that there are breakdowns of communication in feeding information forward. Such a situation exists in property management where the design team seldom supply the property owner with information which can help him maintain the property he has inherited. Although the designer may well have considered maintenance and servicing requirements at the time he specified a product, rarely are his findings communicated to those who assume responsibility for maintaining the property. The preparation of a Maintenance Manual by designers is a convenient form by which the information can be passed on: the Maintenance Manual should be to a building what the owners' handbook is to a car. It should contain as clearly and concisely as possible all the information which is necessary to keep the building in use, clean and in good repair.

The Maintenance Manual should be designed to enable the property manager to organise the repair and maintenance of the building, its services and surroundings effectively and economically; it should enable the occupier to clean the building and operate its services efficiently and reduce losses of time and production; and it should establish another link between the project design team and the client and his maintenance organisation to their mutual benefit.

A Maintenance Manual should contain information on:

(a) General sources of information on the building, its design; construction; contractors; components; contractual arrangements and leases; details of any guarantees affecting components or materials specified, together with their expiry dates.
(b) Instructions on the general maintenance and routine cleaning of the fabric of the building; notes on servicing contracts; manufacturers' instructions, housekeeping requirements.
(c) Drawings and specifications; charts on operating the mechanical and electrical services; periodical routine maintenance patterns; location of meters, recording devices, stopcocks, valves, pipe-runs; characteristics of the building which might otherwise be hidden; special features such as jointing and detailing.
(d) Emergency information giving the names, addresses and telephone numbers of contacts in the event of fire, theft or burglary, or gas, electricity or water failure or leaks.

This link along with the feedback of design/performance reports will provide the communication channels so badly lacking. The faults which tend to be repeated in new designs would be obviated by an increased awareness on the part of design teams of the failures and successes of previous projects. Such reports would also help the building owner to identify his own requirements for both his existing properties and any new buildings he is contemplating erecting. The very task of preparing a Maintenance Manual would focus the designer's attention on his responsibility for the building's performance throughout its life. Maintenance Manuals would save a great deal of the property manager's time at present spent searching for information. The design/performance reports and the manuals can constitute the basis of an information system, the introduction of which would strengthen the procedures at present used for property maintenance management.

Conclusion

An important step in the way ahead is the build-up of knowledge and experience followed by its general dissemination and understanding by all the people who needed to know. Internal information systems and reciprocal exchange services play an important part in achieving this: they promote the willingness to exchange information and share experience.

DISCUSSION

R. A. Diss (London Borough of Waltham Forest)

I am eager to learn techniques which will increase my efficiency as an architect and help me to produce more satisfactory buildings but I do feel that much of this new expertise I am acquiring will be cancelled out if I have to spend my time solving self-imposed problems. I suggest that our life style and expectations are probably two of the biggest factors affecting the performance of buildings today. Perhaps the main question is not 'how?' but 'why?'

J. D. M. Robertson

There are so many real problems to be solved that we do not have time to spend on dealing with insignificant questions however interesting. We should not concentrate on Terotechnology as a concept in its own right only attempt to put it in its correct perspective along side other design considerations. We always have had to involve cost, return, timing, but they are now recognised as something of importance. My paper attempts to cut across the problem of concept and philosophy to describe the contemporary position and to advise only attempting the possible in relation to the extent of our knowledge and our input data terms. Research must be in advance of practice to ensure development but practice must be based on the best information available at the time.

Dr D. Fitzgerald (University of Leeds)

I would like to know the difference between capital expenditure and revenue expenditure (other than borrowing) from the point of view of the government. For my personal expenditure the difference is quite clear, if I buy a motor car perhaps once in five years, that is capital expenditure and I have to save the money or borrow it. But if I were a taxi proprietor with 100 vehicles I might be buying 30 cars a year and I would regard this as revenue expenditure rather than capital. For the government, saving is not possible, it gets money from us all the time on a continuous basis and it spends it on a continuous basis, some on new buildings and some on maintaining existing buildings and we have been discussing whether the government should be spending more on the former to save on the latter. From the point of view of the government it is all money spent, and the aim must be to get a maximum of value from the total spent, whether on building or maintenance. It follows that there is no difference between capital and revenue expenditure, so I find it very difficult to understand why the government imposes capital cost limitations and not real cost limitations, that is costs-in-use limitations.

H. S. Staveley (Martin Staveley & Partners)

Within the Health Service, as far as I am aware, capital is used for new works and revenue is used for maintenance works. There has been a tendency which is not confined to the Health Service to use revenue monies including money which should be used for repairing existing buildings to fund capital alterations and extensions. An older method of division in the Health Service organisation was to describe expenditure under £500 as revenue and over £500 as capital. One definition of capital could be money used for providing something tangible which was not there previously, revenue monies being for consumable expenditure and repairs.

J. D. M. Robertson

This question may seem superbly innocent but it is in fact a question of considerable depth. The concept that the first basic budgetary split is between capital and revenue tends to be taken for granted. It is easily understood in local government because the revenue commitment of a capital expenditure is the loan charges and other charges of upkeep are paid from normal revenue accounts. Mr Bathurst who is engaged in a central government activity might like to comment on the basic philosophy behind the capital revenue division.

P. E. Bathurst (North West Thames Regional Health Authority)

Dr Fitzgerald's point that the money comes from the same source ie taxpayers, or is borrowed from abroad and goes, as far as buildings are concerned, in payments to the construction industry is a deep one. Why do we insist that some expenditure is labelled Revenue and other items Capital. Perhaps it needs an exert in Treasury accounting to give a proper answer but I think it must be along these lines: In general governments do not commit themselves beyond their own lifetime. Therefore they are careful to divide their commitments into long and short term. In the main, Revenue costs (Revenue is a description used in the Department of Health and Social Security, Recurrent Expenditure is used by the Universities Grants Committee) relate to the support of services which had already been approved by previous Parliaments. By and large you continue to pay all the costs necessary to maintain these out of 'revenue' and the only adjustments are made for inflation or changes in those services.

When it comes to a new service Parliament wishes to introduce, this will probably need short term capital investment with consequential Expenditure Revenue. The distinction is really that anything which relates to an existing service will certainly come out of revenue, something which is related to a new service, which is very probably going to need a building or other form of investment, comes out of capital. This in turn produces revenue consequences which have to be taken on board by the succeeding Parliaments. The next question is why are cost limits only related to capital costs? This is really the Government imposing restrictions on expansion on itself. There is no need to set cost limits on existing services because these have already been approved by Parliament, all that is necessary is to provide sufficient to keep these going. If it is necessary to reduce expenditure on existing services this becomes a matter of changing the standard of provision ie perhaps reducing the standard of health care, and certainly this would be considered very carefully by Parliament.

H. S. Staveley (Martin Staveley & Partners)

The picture can get a little confused by the way some people operate or interpret the system. For example, in a hospital project, in order to get the building up within the capital cost limit, the fittings were provided out of revenue monies.

J. D. M. Robertson

Mr Bathurst described how the system works. It works this way for short and long term commitments and is probably successful. But there is a grey area in maintenance, normally called minor works and improvements. Improvements are not maintenance items but sometimes virement is exercised to transfer money tucked away under another budget heading into maintenance, or vice versa. Maintenance budgets themselves are very flexible. There is seldom a precise list of items of work against which there is a priced allocation. There are basic ideas of what might be programmable major works and the rest is an allowance for emergency and day-to-day maintenance. The programming and pre-costing of maintenance work is minimal so there persists a very flexible budgeting situation in the control of the maintenance revenue account. So at maintenance management level the same problem exists as at the design stage—sometimes maintenance is bought out of capital and sometimes capital works out of maintenance funds.

Chapter 13

The Future Role of Terotechnology

H. S. Staveley

HISTORY

Terotechnology as a concept is not new. It is a practice which has been followed by enlightened Governments, prudent companies and thinking individuals right back to the times of the Romans, Greeks and even the ancient Egyptians. It is obvious that such civilisations often considered initial costs in conjunction with good design, life-span and maintenance requirements. It is, therefore, worthy of note that, in spite of centuries of neglect their heritage in many instances is still with us for our instruction and delight.

From time to time in recent years references have been made to a technique which would embrace finance, design, construction, maintenance and eventual disposal. In the building sense, it has often been a battle between initial capital constraints, aesthetic requirements, limited life and the necessity to experiment with novel materials and new methods of construction.

May I take one earlier reference to the concept of Terotechnology just at random? In 1967, the Ministry of Public Building and Works and the Mercury House Journal 'Maintenance Engineering' jointly sponsored a Conference on *Profitable Building Maintenance*. It was at this Conference that Brian Drake in his capacity as Technical Secretary to the Government Committee on Building Maintenance gave an excellent Paper on 'The Economics of Maintenance'. This Paper largely concerned itself with the effect on maintenance economics of decisions taken during the design stage and discussed techniques for bringing to a common basis expenditures which occurred at various points in time. In his introduction Brian Drake said—'In our work for the Committee on Building Maintenance we have found that the subject is virtually as wide as construction itself and in some areas even wider. So that 'maintenance economics' could embrace very wide considerations indeed from the very inception of a design for a building to its eventual demolition.' If one were to substitute the word 'Terotechnology' for the words 'maintenance economics' in that Paper, then one could see the idea beginning to form even eight years ago. At the time of that Conference it was noticeable that there had been very little literature published on the subject of building maintenance and, as Brian Drake stated in his Paper, there was even less conclusive data on the subject of inception right through to eventual demolition.

Reverting back to the broader aspect of Terotechnology in relation to both engineering and building, in April 1970 the then Ministry of Technology published a Report by The Working Party on Maintenance Engineering which showed that British manufacturing industry wasted large sums of money each year because of ineffective and badly organised maintenance. The Working Party were driven inescapably to the conclusion that improvements in life-cycle performance could only be achieved by the co-ordinated application of the several disciplines which had not previously been brought together in any such way.

The Committee for Terotechnology was, therefore, formed in April 1970 to advise the Government on the best methods to employ in order to ensure the establishment of Terotechnology practice in the commercial and industrial life of this country. This

Committee is still in existence and now works within the structure of the Committee for Industrial Technologies.

Meanwhile, on a parallel plane, the Committee on Building Maintenance had been appointed in August 1965 by the then Minister of Public Building and Works as a Standing Committee whose terms of reference included the continuous review of problems involved in the maintenance of buildings, including the relationship between design and maintenance, and to make recommendations accordingly. The final Report of this Committee was published subsequently in 1972 by The Department of the Environment and it is apparent that the conclusions reached included many which were very similar in principle to those reached by the Working Party on Maintenance Engineering. For example, Chapter 3 of the Report of the Committee on Building Maintenance states—'It is clearly important that the correct balance for the individual unit of the economy in general is struck between expenditure on new works with that on maintenance and that decisions are based on the appropriate economic and other relevant criteria. It seems likely that in many cases this correct balance is not struck and that inadequately based decisions are made.'

It is also mentioned that there should be an appraisal of alternative building and design decisions as for any other investment and that these should take into account all the costs and benefits that are likely to arise during the life of the building, including the potential maintenance costs of the design. At the same time it is pointed out that buildings cannot be viewed in the same light as plant and machinery and that the great majority of buildings provide shelter for people whose wants and needs must be understood. In effect, the design and maintenance of buildings involves a social system in which not only maintenance management but the designers, the operatives, the building owners and the occupants all play active parts and often have conflicting interests.

Although it is apparent that there had been little, if any, connection between the research carried out by the Working Party on Maintenance Engineering and that carried out by the Government Committee on Building Maintenance, the relationship between the principles involved has by now become obvious. During November 1973, the Committee for Terotechnology held a One-Day Forum on Terotechnology to which, *inter alia,* members of most of the professions associated with the Construction Industry were invited. Until that time, the application of this supposedly new technique had spread only as far as the engineering world and had been confined almost totally to plant and machinery. Although, possibly through lack of interest, professional representation from the Construction Industry was severely limited at this Forum, during the course of the discussions it became quite obvious that the principles involved were equally applicable to the Construction Industry generally. The Department of Industry were swift to acknowledge this fact and as a result, they sponsored and encouraged the formation of the Building Terotechnology Group, at the same time collaborating closely with the Department of the Environment. Consequently this Group was set up under the joint sponsorship of both Departments. This close liaison between the two Government Departments has now resulted in a Group with a membership representative of the design, maintenance, engineering, construction, financial and client aspects of all matters associated with buildings and structures.

The need for an overall consideration of life-cycle costs and the necessity for better liaison between client organisations, design, construction and future maintenance have been further emphasised in the Report of the Woodbine-Parish Committee which was a Government Committee of Inquiry into hospital building maintenance and minor capital works. Furthermore, that part of the Terotechnology process which is concerned with design in relation to future maintenance would also include the consideration of design failures and these, again, have been stressed by recent events. This is particularly so in the adoption of The Defective Premises Act 1972 which imposes duties in connection with the provision of dwellings and lays down the liability for injury or damage caused to persons through defects in the state of premises. Although the scope of this Act is limited to new dwelling houses, it is most certainly a step in the right direction and emphasises the duty of care which is required where work of construction, repair, maintenance, demolition or any other work is done on or in relation to premises. Further increases in this acknowledgement of the responsibility of a designer to his client in constructional matters followed and, in particular, one could mention the case of Sutcliffe v Thakrah and others, an Appeal where the House of Lords over-ruled the long-held theory that an Architect employed under the Standard RIBA Contract acts as an Arbitrator and is, therefore, immune from action in negligence at the suit of a building owner when he has issued interim Certificates covering defective work.

It was now becoming more and more apparent that the division of the various phases of a building's existence could no longer be kept apart in distinct and separate compartments from both a practical and professional view and that, right from the outset, an overall view on both cost and use basis must be employed. Again, as if to reinforce this accelerating trend, the Royal Institution of Chartered Surveyors recently reported back to the Department of Environment on the subject of Recommendation 26 of the Report of the Committee on Building Maintenance. This Recommendation read—'Statutory or voluntary arrangements should be developed to provide an extended responsibility by designers and contractors for defects in new buildings'.

Incorporated in this Report were some alarming statistics obtained from The Building Research Advisory Service in which detailed investigations were made into 500 defective buildings over the period 1970 to 1974. Although it is appreciated that these problems were not necessarily representative of the generality of building defects, they gave an obvious pointer to current trends and, as such, gave cause for alarm. For example, it was found that 58% of the sample had defects attributable to faulty design, 35% attributable to faulty execution, 12% to faulty materials or propriety systems and 11% to unexpected user requirements.

Briefly, the recommendations of the RICS Report included a better appreciation of materials in the training of designers; greater emphasis on maintenance in the training of designers, surveyors and operatives; better feed-back of performance information to the designer and building owner; improved training for building operatives and craftsmen together with incentives and recognised pay differentials; liability of builders and designers for latent defects to successors in title and, lastly, the examination of all designs from a maintenance viewpoint.

The obvious need for improved co-operation between client/finance/design/ construction/maintenance/user has, therefore, been firmly established during the last ten years and this, after all, is what the concept of Terotechnology is all about.

CURRENT PRACTICE:

There are undoubtedly many who dislike intensely the word 'Terotechnology'. However, nobody has yet been able to provide a constructive alternative in the form of a single word and one would have thought that lengthy phrases describing this function are to be deplored. The word itself is now in international use, at any rate throughout Europe, the Americas, Japan and Australasia. Indeed, the Japanese have already published a comprehensive text book on the subject of Terotechnology, this having been printed in both Japanese and English! The definition for Terotechnology appears in the recently revised British Standards Institution Glossary of Maintenance Terms in Terotechnology, BS 3811 1974 and many will already be familiar with this publication.

However, the Building Terotechnology Group have revised this definition as far as application to the construction industry is concerned, to read—'a combination of management, financial, engineering and other practices applied to physical assets in pursuit of economic life-cycle costs. In the construction sense this practice is concerned with the design and specification for reliability and maintainability of buildings and structures *including their associated plant, machinery and equipment; with their installation, commissioning, maintenance, modification and replacement and with the feed-back of information on design, performance and cost.'* As some may find it a somewhat indigestible and lengthy description of this function, perhaps it could be easier defined as either *'resource management', 'cost of ownership', 'cradle to grave management', 'physical assets management'* or *'life-long care'*.

A word now about the multi-disciplinary Building Terotechnology Group. When this Group was formed, the purpose was not only to bind together those professions, often rival, who are concerned with a building at any one particular phase of its life, but also all the interested parties who are concerned with a building through the whole of its life-cycle from inception to demolition. Thus, we have a Group on which are represented the Government, client interests, finance and accounting, design, construction, user, operation, maintenance and even demolition and eventual disposal. For example, it has not always been appreciated in some circles that client and user are seldom one and the same person. Their requirements could well be vastly different and certainly the demands made upon a building during the course of its life can vary considerably from user to user and in each case these could differ wildly from the requirements of the original client for whom the building was constructed.

In addition to the furtherance of liaison between all parties concerned with buildings and structures through all phases of their lives, the Group also acts as a catalyst for research and a watchdog to prevent unnecessary duplication of valuable research resources between the various disciplines and interests involved. The Group's terms of reference are drafted with the intention of promoting improved understanding and co-operation between the user, designer, contractor, operational and maintenance staff together with other professional and technical advisors concerned with buildings, structures, sites and services for achieving the most effective use of resources for the built environment. One of the objectives of the Building Terotechnology Group (BTG) is that of developing a more informed approach to the planning, design and construction of buildings with a view to increasing efficiency and ease of maintenance during their life-cycle.

At this point it must be stressed that one must not make the mistake of assuming the words 'Terotechnology' and 'Maintenance' to be synonymous. Many regard maintenance as being the axle on which rotates the wheel of life-cycle costing. It is, however, quite logical to regard maintenance as being a central feature of Terotechnology in that it occupies the mid-way position between construction and eventual disposal. Terotechnology only serves to emphasise that maintenance should be eliminated whenever this can be justified and when this is not possible, maintenance should be minimised.

One point which should be stressed is that there is no such person as a 'Terotechnologist' and never will be. Terotechnology is a multi-disciplinary subject and is a combination of various skills which have been assembled and co-ordinated managerially in order to carry out a particular task. As has been said in the earlier part of this paper, the concept is not new but there is, at the present time, a very great need to emphasise to everyone the complete and unassailable necessity for considering the total life-cycle cost implications of every single design.

Life-cycle costs include the costs associated with acquiring, using, caring for and disposing of physical assets including feasibility studies, research development, design, production, maintenance, replacement and disposal as well as the support, training and operating costs generated by the acquisition, use, maintenance and replacement of permanent physical assets. This rather cumbersome and mind-bending description boils down to the fact that we are concerned with the total ownership cost of physical assets including costs of operation which are also part of the total life-cycle cost, although frequently neglected, particularly at the design stage.

The urgency of this approach towards the ownership of buildings and structures has recently received an added impetus by the attitude of the OPEC countries as far as major oil fuel costs are concerned, together with the embarrassment of the increasing alternative fuel costs in other fields, this mainly being due to excessive wage increases, poor labour relations, high transport costs and the rapidly increasing scarcity of all currently usable natural energy resources. It has, therefore, come as no surprise to find that the conservation of energy and resources involving the elimination of all unnecessary waste should form a very large part of the BTG's programme and loom large in the Group's list of priorities. In fact, it was found that, so rapidly was one becoming overtaken by events during last year, that it became necessary to form a Working Party on Energy Resources out of the BTG membership, the terms of reference of which were—'to consider current theory and practice for the optimum use of energy resources in buildings, with special emphasis on the problem of existing buildings and to make recommendations to the Government, building owners and users and to the various disciplines within the building industry on ways in which the application of Terotechnology or its principles will be beneficial.'

This Working Party are taking note of energy conservation work elsewhere and not only seeking to encourage further research, but also endeavouring to ensure that research work is not duplicated amongst other bodies. They are to take into account the effects of legislation, fiscal and taxation policies, building cost control methods, energy budgets and systems of tenure and also the need to develop economic criteria for standards of insulation above the legal minimum. As the pursuit of energy conservation is at the moment highly fashionable, it is therefore incumbent upon the BTG Working Party to ensure that human resources are not wasted nor confusion caused by a proliferation of research and reports, some of which are often contradictory.

Conservation of energy was the main theme of the 1975 RICS Edinburgh Conference and it was realised at that time that, as far as new buildings were concerned, decisions at design stage established the pattern of energy consumption and use which was to be carried forward for many years into the future. It was even suggested that multi-skilled design teams capable of investigating alternative techniques available in the design of new buildings to minimise energy consumption would be required in the future. Declaration of energy targets would then form part of the planning process and, indeed, possibly even be included in the requirements for Planning Approval for a new building. Again, there should be a reappraisal of financial accounting methods for the use of energy in order to obtain a real incentive for energy saving. In this, of course, one has to labour against the present taxation system which is not oriented towards such incentives.

As far as building services are concerned, Heinze Steinacher, the well known European Director of Machinery and Systems Technology, has stated that at the moment in the UK the cost of petroleum imports in 1974 amounted to more than £2,000 million. With 25% of this oil used for space heating and for the generation of electricity for air conditioning, the total expenditure for this user segment is £500 million. The waste-rate of heating and air conditioning systems is about at present about 26% or in terms of money about £130 million per year. There is no doubt that at least 50% of this waste could be eliminated which would result in an annual saving of £65 million. This saving would be dependent upon good system design and the optimisation of project requirements.

This optimisation would undoubtedly involve compromise because of, for example, the conflict between initial costs versus operating costs and short-term versus long-term considerations. Heinze Steinacher warns against 'legislative actions hastily and ignorantly prepared by well meaning but ill informed politicians', but says that at the same time one must be ready to assist them in the preparation of meaningful legislation for energy conservation. This can only be done by people with real experience in this field and the co-ordination of such effort for advisory purposes is where the services of the BTG Working Party are most needed.

Terotechnology has for many years been well established in the fields of aviation and defence where complete efficiency and reliability are absolutely essential for the preservation of life and costs are very often thought to be a secondary consideration. However, in the present economic climate costs necessarily have to be very closely considered and we thus have the concept of Terotechnology practiced by establishing a very close relationship between reliability and costs.

One factor which has received far too little consideration in the past has been pointed out by Professor Markus of The Department of Architecture at The University of Strathclyde during a discussion group at the *Terotechnology for Better Resource Management* Conference held in London during April 1975. Professor Markus expounded the theory that in plant, machinery and vehicles, as in buildings, the ideal would be for all components to wear out at the same time, thus ensuring a replacement-free article together with a consequent lowering of maintenance costs. This would then enable the life-span of the machine or building to be evaluated far more accurately than it can be at the present moment when so much is attributable to the standard of the maintenance which has been carried out during the life-cycle of that item.

In this connection, therefore, Terotechnology is concerned with the optimisation of the whole life-cycle of the physical asset concerned. This, however, is where engineering and construction begin to diverge as one then has to consider that the life of buildings normally far exceeds that of mechanical equipment due to the comparatively rapid obsolescence of the latter together with the absence of moving parts in buildings and the stringent conditions imposed on standards of construction by the Building Regulations, planning constraints and the effect of other legislation such as The Fire Precautions Act, Defective Premises Act and similar statutory requirements. These considerations may have a profound effect on both the initial and running costs of a structure outweighing purely financial and practical aspects.

Before leaving present events and looking more to the future, mention must be made of the increasing interest which has been taken concerning design failures in buildings. The Building Maintenance Cost Information Service, for example, have a Design/Performance Feed-back Section in which defects in construction and finishes are reported, commented upon and alternatives suggested. Information is provided by contributors to the BMCIS Service but there is occasionally a tendency for the defects which are reported to be of minor significance and hardly worth the printing. However, there have been some invaluable experiences published and interesting suggested alternatives provided in the past and this is undoubtedly a service well worth supporting. In addition, a series on 'Design Failures in Buildings' was printed in *Building* in the form of Building Failure Sheets. The principle of these Failure Sheets was similar to the Design/Performance data given in the BMCIS Service, but perhaps rather more detailed and concerning perhaps more significant elements in the building's construction. These Building Failure Sheets have been published by George Godwin Ltd in the form of two volumes (First and Second Series) and would well repay a close study by designers. Each failure is dealt with under two broad headings; *Why did it fail?* and *How it could have been avoided.*

As the question of feed-back of information from user/operatives/maintenance back to drawing board has proved extremely difficult in practice, mention has now been made that the employment of standard details in isometric form as well as plan, section and elevation should be prepared by the designer and used in the construction process down as far as the craftsman and building operatives. These would refer to standard details of elements, components and construction generally. Although this trend is thought by some architects perhaps to inhibit original design, it receives a great deal of support from many well known figures in the architectural profession. In this connection, although strictly not standard details, the National House-Building Council publish a Registered House Builders Site Manual which involves excellent and simply illustrated information on what features constitute good construction in a dwelling house, also illustrating the serious effects of defective workmanship and defective materials.

FUTURE TRENDS

Past-President of the **RIBA** Alex Gordon speaking at HRH Duke of Edinburgh's Study Conference held at Cardiff in July 1974 stated, 'Our predecessors left us with a stock of buildings which have, in general, been pretty adaptable and served for a long time. One suspects that many of today's buildings are only going to be really suitable for the functions for which they were designed for a comparatively short time and we have embarked on the mad escapade of tearing down and rebuilding much of what was built in the last century. We hear increasing talk of re-cycling materials such as metal and paper; we must now begin to talk more about the re-cycling of buildings. From now on the best service an architect can, in some cases, render to his client (and to future generations) is to advise that a new building is not required and the old one can be rejuvinated. This is already happening increasingly in the housing field.' In other words, not only is it essential for us to have an improved understanding between architects, maintenance manager and engineer, but also between all three of them and the client, including the client's financial adviser. This series of dialogues is vital and the whole application of the terotechnology concept hinges on the very close co-operation of all these parties.

The 'Long life, loose fit, low energy' view advocated by Alex Gordon will most certainly become a necessity in the future due to the high cost of fuel, labour, materials and the ever diminishing supply of natural resources. At the brief stage, the client will then not only have to consider his own immediate requirements but should also weigh the possible requirements of any future user who may wish to change the purpose for which the building was originally intended.

One typical example illustrating the wastefulness of a purely single-purpose design is that of the usual multi-storey, pre-stressed concrete, city centre car park. Such a structure is almost completely unsuitable for conversion for any other purpose and is thus entirely dependent upon the continuance of its present use. It is a well known fact that, in the present economic climate and with possible future planning constraints, the city centre multi-storey car park could become superfluous. Some cities have already seriously considered the elimination of private vehicles from city centres (Bristol being a recent example) and there is every indication that such measures are likely to take place in the future when traffic conditions become impossible in the centre of large conurbations.

If this were to happen, we should then be left with a great number of large buildings, all of which possibly occupy the busiest and most valuable sites in city centres and which cannot be used for any other purpose than that for which they were originally designed. Without any further complications, buildings of such a size in such situations would be extremely difficult to demolish, but couple this with the fact that these buildings are often composed of pre-stressed concrete which would make the structure dangerous to demolish in such surroundings, then we are left with a considerable legacy of wasted effort and, in some cases, even potential 'time bombs'.

Not only must the designer in the future, and more importantly the client, consider the flexibility of his building at a very early stage but he must also consider the life-span of the building in connection with his requirements. Once again, the ideal would be to make the building as permanent and maintenance-free as possible, using the longest lived components consistent with the dictates of obsolescence. As I have already pointed out, it has been said in some quarters that permanence in a building inhibits progress in original design, but events are undoubtedly moving rapidly in a different direction and undoubtedly permanence in buildings will have to be a factor of the utmost priority to avoid unnecessary waste. The architect's skills then may well have to be deployed into adaptation and rehabilitation works, concentrating on making the best of all the existing stock of properties at our disposal.

Therefore, let us say that at briefing stage the client ought to receive the best possible advice, not only from his designer or architect but also from his financial adviser and that this brief should be based on the results of very close co-operation between these three interests. At the same time, either the designer should receive a much better training in the art of building maintenance or alternatively he should be in a position to consult others who have greater experience in the maintenance field before his design leaves the drawing board stage. One feature which could well become more common practice in the future in a *desk appraisal* of an architect's design by a maintenance surveyor at final design stage. Pilot schemes based on this principle have in the past been carried out with success, in that large structures which could have proved difficult and expensive from the point of view of future maintenance have been amended while still at final design stage.

Such a service also offers an invaluable check on the architect's drawings from a construction point of view in addition to that of the future maintenance implications. Perhaps the incorporation of maintenance surveyors within architect's practices may well have the desired effect and it is to be hoped that this aspect of the terotechnology process will be fostered in the future. If the brief dictates that as maintenance-free a design as possible should be employed but on the other hand limits the capital allocation for this purpose, then it may well be necessary to consult the maintenance side of the professions in order to seek their advice in any event.

At the time of the Profitable Building Maintenance Conference held in London during 1967, planned maintenance was very much in its infancy and little, if anything at all, had been written or spoken on the subject. Since that date a great deal about planned maintenance has been said and published and one now has no need to elaborate on this subject, save to say that it also follows on as part of the total terotechnology process. The experiences obtained by the maintenance manager as a result of operating a planned maintenance schemes must somehow be fed back to the designer on the drawing board in order to facilitate this consideration of design as related to future maintenance. The alternative to such a form of feed-back will have to be the greater reliance placed upon standard details of proved construction as previously described. This need not necessarily have a totally inhibiting effect on the designer but could well eliminate a great deal of poor design, which is far more important than the mere search for novel effects.

The interpretation of the present trends as seen in Government policy, difficult labour relations, weak management and strong Trade Unionism, lead one to believe that any structure which will absorb the minimum of labour costs during its lifetime is to be encouraged. The ideal to be aimed at would be a permanent maintenance-free structure and although this ideal may be impossible to conceive in its entirety, one should always aim in this direction and in that way ensure that at least some progress is made towards the ultimate object, ie a permanent maintenance-free structure. There should be no reason to encourage such a labour-intensive industry as building maintenance except for, perhaps, purely political reasons.

As Professor Markus implies, it is of little use introducing components into a building which have the result of rendering that building obsolete when other components incorporated in the structure still have many effective years to run. We have all seen the tremendous amount of waste which results from the demolition of a building or structure and how very often many of the materials which should be recoverable and are of immense value are simply bulldozed into the ground or carted away to tip.

Without going into too much detail, it is essential during this phase of the terotechnology process, always to bear in mind the facts that scarce and sometimes valuable building materials and components which are disposed of in this cavalier fashion are more often than not irreplaceable in that materials and components of a similar quality cannot any longer be found. Sociologically, there is also the problem of waste which has resulted from the summary disposal of an end product in which is embodied so much human ingenuity, effort and material resources. Much greater thought must, therefore, be given to the salvage and re-use of timber, slate, tiles, steel, bricks, stones, blocks and components such as windows, trusses, glass, and fittings of all kinds. At design stage, therefore, flexibility in the use of a building to avoid unnecessary demolition as well as the salvage of components and materials when such a demolition becomes unavoidable must be closely considered as a part of the terotechnology process. Not only must this include the physical aspects concerned with the termination of a building's life but also the cost implications of demolition and possible re-building to serve an alternative use.

As many clients are inarticulate in the design sense, these facts will have to be pointed out to them by the architects at briefing stage and they must consequently be made to see that available capital and accommodation requirements are not the sole criteria governing a 'best buy' building. With the future trends of increasing labour costs and material shortages, the maintainability of a design will possibly have to take a far greater priority over purely aesthetic considerations than has sometimes been the case in the past. These inarticulate tendencies on the part of the client have often resulted in a further lack of communication which has sometimes led to complaints concerning defects in a new design, and perhaps in the construction of a project, being circulated from the client to his friends, his club, business associates and perhaps even eventually into the media without the architect having the slightest knowledge that his design was under fire until, perhaps, the criticism reaches a peak at which stage his reputation may suffer.

A future trend could well be for building owners and all professions concerned with the design, construction and operation of buildings to be able to ventilate their opinions and complaints through Regional Building Terotechnology Centres which could well be set up under the auspices of the Building Terotechnology Group and act as effective 'clearing houses' for such matters. It could be that such units could be established at the existing provincial Building Centres and, as there is now established a National Terotechnology Centre under the able direction of Dennis Parkes, no doubt assistance in the formation of provincial units could also be given from this source. The National Terotechnology Centre acts as a national focal point for terotechnology and provides free basic information and advisory services in both engineering and building fields.

Reverting back to the life of the building, there is one factor concerning environmental design which perhaps has not received sufficient consideration in the past. I refer to 'design quality' and its effect on the life of a building. There is such an emotion as strong affection for a structure and this will undoubtedly prolong that structure's life, as witness a number of churches and historic buildings which no matter how badly maintained or how incompatible with modern needs, still endear themselves to the general public. Consequently, when demolition and re-development are suggested there is a tremendous outcry and very often a building is retained which serves no practical need or, in any event, does not fulfil a need in the most effective manner. This, however, is no bad thing and the virtue of good manners in design, as distinct from the sole motivation of a 'desire to be different' should be fostered.

Who then are we to say what will constitute good aesthetic design in 200 years time? A matter of forty years ago we were brought up to appreciate that Victorian and Edwardian buildings were the ultimate in poor taste and ugliness but we now find many of these buildings scheduled as being of historic interest, late 19th century development designated as conservation areas and the aesthetic delights of such construction being eulogized and romanticised by Sir Hugh Casson and John Betjeman! The works of Le Corbusier, Art Nouveau and even Art Deco are also considered worth considerable appreciation and, as tastes obviously change from decade to decade, we must, therefore, increasingly consider sound construction, functional planning and maintainability to be the essential ingredients of present day design as these are the qualities which are never changing. In the future we could do worse than to follow good manners, good construction and good planning rather than be influenced by fashions in shapes and finishes.

Continuing, this time on the theme of life-cycle costs, accurate figures are virtually impossible to obtain as yet and the BTG hope to foster case studies on this very subject in the near future. Although the purchaser of a motor car, for example, is largely influenced by the impact of life-cycle costs and the battle is between maintainability and the expendability, this consideration is often neglected entirely or subordinated to other considerations in the case of building design. Recent figures for GLC house building indicated that the policy of demolition and replacement which was followed from 1965 to 1971 as far as their housing stock were concerned, may well have resulted in a great deal of unnecessary expenditure which could have been obviated had the demolished stock been retained for perhaps a further fifty years, rehabilitated and brought up to 1971 standards.

A future trend which could be of the greatest importance is the retention and rehabilitation or conversion of existing building stocks, be they housing, hospital, factory, office or entertainment. In the case of the GLC figures one could say that the application of the principles of terotechnology would have led to a different result in that a cost comparison between maintenance, rehabilitation and further maintenance on the one hand and demolition, re-building and maintenance on the other would have been made after discussing the relative cost merits of both courses before the 'point of no return' had been reached. If this had been done, there is no doubt that savings could have been made which could then have released a considerable amount of monies from which the stock of houses could have been increased, fiscal system permitting.

If, in the future there is less labour concentration on maintenance then, of course, there is likely to be considerably less employment available in that field. Should this lead to Union objections, it may well be pointed out that such experienced labour could better be used in the rehabilitation of existing structures and that this is a process which would be likely to occupy such labour for many years to come. In other words, the future need is not to 'maintain' ie keep the building in its existing state until such time as it becomes obsolete, but rather to 'improve' so that the building then keeps up with current requirements and obsolescence is avoided.

The sociological consequences of the terotechnology concept are not always fully apparent at first sight. Sir Ove Arup at the inaugural meeting of the Building Services Engineering Society on the 26th October 1972 said 'During the war we were told to ask ourselves whether our journey was really necessary in view of the need to save resources for the war effort. We have a war on now and we would do well to ask the same question—we must pay more attention to the first problem. We must take a critical look at the brief, make it more comprehensive, we must look beyond the narrow object and ask ourselves what would be the ecological consequences. What about the working conditions for those who carry out the work, including their spiritual well-being—will the work provide useful employment or cause unemployment, perhaps in other countries? What effect has it on other industries? What is the cost in scarce resources? We must ask ourselves what would happen if everybody else did what we do. Would that serve humanity?' This quotation sums up the need to look far beyond the narrow field of immediate needs.

Conclusions

In conclusion, the practice of terotechnology requires no fundamentally new disciplines nor yet the creation of a new hierarchy of specialists. It is essentially a means of formalizing and facilitating collaboration between professions, clients and finance in their own fields and at all levels in the organisation. Although there will need to be a broadening of the education and training of many of those concerned because of the diverse nature of the subject, it will not be necessary or desirable to create a new class of specialists calling themselves 'Terotechnologists'.

Because terotechnology is based on existing knowledge and disciplines, the main requirements for its implementation will be:

(1) The acceptance of a multi-disciplinary approach to the running of an enterprise (for example, by the inclusion of an accountant in the design team or by the maintenance manager being consulted over design and construction techniques and schedules).

(2) An appreciation of the function and workings of an enterprise and the role of a particular job or discipline within the framework.

(3) A broadening of the education and training required for all those responsible for the operation of an enterprise.

(4) An appreciation of the whole life-cycle requirements of a physical asset.

By these means we hope that there will be found a better understanding of each others problems, that destructive criticism and unnecessary competition between allied trades and professions may be reduced, that existing structures, materials and components should not be wantonly destroyed without good reason and that the total cost consideration of every new structure is the result of a happy marriage between practical design and sound building construction.

The practice of these principles will, for all concerned, mean a better knowledge of the factors which affect buildings, building services and structures through their complete life-cycle and should then result in improvements in design, safety and efficiency; not only will this widen the scope of professional expertise but will confer increasingly greater benefits on the built environment both now and in future years.

DISCUSSION

R. F. King (British Airways)

I would like Mr Staveley to expand on the cost comparison he gave in his presentation between the Greater London Council's new Island site and the old County Hall. The cleaning costs showed about £97/m² difference, is this due to the new building being air conditioned with cleaner filtered air?

We produced a clean air suite for aircraft instrumentation operators some years ago. Superficial enquiry on the medical effect of the low dust levels indicated a reduction in the average number of colds amongst clean room staff. I wonder if there is other evidence of the effect of air conditioning on the health of staff, whether beneficial or otherwise.

H. S. Staveley

I think the principle reason for the decrease in cleaning costs is probably the thought that has been given to the type of materials, the shape of them and the way in which they are used. For example, unnecessary projections have been avoided and there are no column radiators. There is also a piped vacuum system for vacuum cleaning.

It may be of interest that the Island block was designed originally to incorporate a heat reclaim system in order to use the heat produced by the occupants of the building for heating domestic hot water. However initial capital cost was too high and domestic hot water is now produced by gas fired boilers with the result that energy is being used to cool the building via the air conditioning plant and more energy is used to heat the water. Terotechnology did not succeed here because initial capital resources were not available to arrive at the best long term conclusion.

A. F. C. Sherratt (Thames Polytechnic)

In comparing the Island site and the County Hall it must be realised that the usage of the two buildings is very different. The Island site is predominantly occupied by full-time staff whereas County Hall has a large proportion of its area used by Council members and visitors attending committees etc. Interpretation of the statistics is important, a cost of cleaning per member of staff could be totally distorted at County Hall by the large area not used by staff.

H. S. Staveley

It would be a false assumption to take these figures too literally because in the Old County Hall the figures do not include for the people moving in and out. So County Hall has a lower occupancy than the new Island block. As against that, conditions in County Hall where people *are* working, are more crowded than they are in the new Island block which is an open plan building. I think therefore it is reasonable to compare the cleaning figures.

H. R. McCallum (Scottish Development Department)

The differences in cleaning costs for the two properities, ie GLC County Hall £28.50 per head of staff and GLC Island block, Waterloo £11.50 per head of staff, if taken at face value is a dramatic one and provides a powerful case for the exponents of Terotechnology. However, is that difference entirely due to thorough pre-contract planning taking account of all identifiable influences on, among other things, the cost of cleaning? For example:

1. What is the ratio of staff to floor area for each building?
2. Is there a different standard of cleaning required?
3. Are cleaning personnel of the same status, ie contract or direct employees?

H. S. Staveley	One factor which has to be taken into consideration is the fact that between 1971 and 1975 staff wages have doubled and electricity charges have risen by 64% which will of necessity alter comparison figures between the old County Hall and the Island block. Whilst I do not have the exact ratio of staff to floor area for each building it is possible that the old GLC County Hall may have been over-occupied and certainly accessible transiently to more members of the public. In that sense, therefore, the ratio of occupancy to floor area in the old County Hall is likely to be higher than in the new Island block.

Again, when the Island block was being designed there was very close cooperation between the architects and service engineers project team and with the two departments' maintenance staff. At this time cleaning as well as maintenance requirements and characteristics were taken into account and about six months before completion of the service contract a maintenance engineer was appointed, the selection board consisting of both design and maintenance staff. Since that time there has been continuous feedback to the Council's in-house design staff and contractors, not only on building and planned maintenance but also on cleaning matters. The standard of cleaning is, therefore, in the Island block likely to be higher than that in the County Hall due to the fundamental design and consideration which has been given to this aspect.

It is understood that the cleaning staff for both the County Hall and the Island block are of the same or similar status.

One point I would like to make is that the upper floors, which comprise the main offices, are successively stepped back to screen the offices not only from traffic noise but also from dirt and to provide easy access for cleaning and maintenance. The air conditioning system which is employed also results in the reduction of the dust content in the air.

R. J. Wilde (Westminster City Council)

I have for a number of years worked in and operated both air conditioned and traditional buildings. I do not think there is much variation in the number and type of complaints from staff in either environment. You will always have the office spokesman. In air conditioned buildings I think you have to educate the staff in their use of the office and in particular in the use of blinds and the planners in their office arrangements. We ask staff working on the East side of Westminster City Hall to draw their blinds before they leave at night but only have a 25% response. The result is that when they arrive at 09.00h the eastern side is already "cooking" and the refrigeration plant working close to capacity.

We have also had problems in the open plan at City Hall with the indiscriminate use of partitions to produce small enclosed office areas suffering extremes of temperature because of the restricted air circulation. The gang switching of floor lighting is also a handicap to the use of limited lighting.

There is a tendency for staff to block off the window-sill outlets with books etc. in the misguided impression that they are 'stopping the draught'. With an induction system of course they are restricting the heat output in the winter and they wonder why the office is colder.

H. S. Staveley

The GLC Island block has experienced similar problems although in this case the blinds are external and have not been without difficulty in operation. I think that in a fairly exposed situation it is not very good practice to have anything fragile externally.

A. J. Colledge (Department of the Environment)

The blinds at the Greater London Council Island site are reported to be torn and tatty looking. At the Department of the Environment we have carried out experiments recently on external blinds for the Tate Gallery extension. These experiments extended over a period and with the vagaries of the UK weather we had serious troubles producing a satisfactory blind to help the thermal and lighting problems. More investigation needs to be done but a successful result could be rewarding. There are many examples in Switzerland and Austria but so far a viable system has not been developed that can successfully cope with our particular climatic conditions.

T. Smith (Steensen Varming Mulcahy & Partners)

Mr Staveley in his paper said that it is necessary for materials in a building design to relate sympathetically to each other in durability. I thought it might amuse the audience to know of a somewhat peculiar building with which we are involved at present and which makes certain demands in this sense that are unusual. It is a temple in the Middle East of dimension greater than St Paul's Cathedral in London. It is fully air conditioned. To equate the dynamic engineering services systems to the building in term of durability was impossible. The building in its natural stone may stand for several hundred years, the pumps and boilers and fans will inevitably have a lesser life.

We approached the question of sympathy between structure and services by choosing materials for services of normal life but designing them into the building in a replaceable form. In these terms the services can almost be described as 'throw away components'. This philosophy of design in fact helped to determine the nature of the engineering services to be installed. In this particular project Palaeo Engineering and Terotechnology became uncommonly related.

Chapter 14

Summary

P. C. Venning

In this conclusion I would like to draw out a number of points from the chapters all of which have made a significant contribution to this important subject.

Peter Bathurst has extended the concept of building design appraisal beyond running and maintenance costs into the financial implications of a building's actual function and use. Two points that emerged which are particularly interesting are, how the form of building tenure can affect the way it is maintained and the other was the plea for effective feedback of cost-in-use data.

The case studies provide, in a very digestible form, an insight into part of the range of building appointment we have to face and a reminder of the basic principles of good design. (I am sure we all envy Building Design Partnership for having found the ideal client we have all been looking for.) The measurement of services was an interesting point for discussion. Quantity surveyors may like to know that my own practice has also adopted the policy of asking its own engineers to produce engineering quantities.

Nicholas Thompson has given us an insight into the obvious deep satisfaction of the creative designer and I thoroughly enjoyed this chapter which was entirely appropriate to the subject by the way it added perspective. The discussion following Mr Drakes paper produced some important points. Personally, I am very glad to know the equivalent current cost of York Minster and I am suitably nervous of the possibility of relating building regulations to a declared heating system. I would like to congratulate Mr Carver and I would thank him for recognising that the procedures he advocated do require design time. I shall take a long time to recover from Professor Burberry's comment to the effect that as a quantity surveyor I belong to a 19th century profession.

May I thank Roger Flanagan in particular for his quotation from the American Manager "I do not care what you do but I want no surprises" and Mr Staveley for the definition of Terotechnology I have been waiting to hear for some time. Running through the papers and discussion were a number of threads that are not new but worth repeating; the question of information feed-back and its standardisation; the need for teamwork and effective leadership in design and construction: education and training policy for the building industry and the building professions (perhaps joint education and closer links between practice and teaching).

The philosophical conflict in the conference and this subsequent publication could, perhaps inadequately, be summarised by "the Drake school" versus "the Bathurst school'. I must in all honesty admit to inclining towards Mr Drake. There are so many assumed factors in the arithmetic of widening building design cost appraisal, it is like solving a set of simultaneous equations with the same number of unknowns. It can be done but takes a long time. By the time it is finished, the builder has probably completed the building or the client has decided not to build it. Nevertheless, this analogy is in no way intended to under-estimate the value of the material contained herein. I know very well, however, that even after assimilating it I shall return to the office to an in-tray containing; an approximate estimate to be done in too short a time; a problem around the Housing Cost Yardstick; and to a contractor's claim, but I hope to be a slightly better quantity surveyor tackling these problems as a result.

Index